SOLIDWORKS 2019
Advanced Techniques

Advanced Level Tutorials

Mastering Parts, Surfaces, Sheet Metal, SimulationXpress, Top-Down Assemblies, Core - Cavity Molds & Repair Errors

Paul Tran, CSWE, CSWI

SDC
Publications

SDC Publications
P.O. Box 1334
Mission, KS 66222
913-262-2664
www.SDCpublications.com
Publisher: Stephen Schroff

ISBN-13: 978-1-63057-224-2
ISBN-10: 1-63057-224-1

Printed and bound in the United States of America.

Acknowledgments

Thanks as always to my wife Vivian and my daughter Lani for always being there and providing support and honest feedback on all the chapters in the textbook.

I would like to give a special thanks to Karla Werner for her editing and corrections. Additionally, thanks to Kevin Douglas, Dave Worcester and Peter Douglas for writing the forewords.

I also have to thank SDC Publications and the staff for its continuing encouragement and support for this edition of **SOLIDWORKS 2019 Advanced Techniques**. Thanks also to Tyler Bryant for putting together such a beautiful cover design.

Finally, I would like to thank you, our readers, for your continued support. It is with your consistent feedback that we were able to create the lessons and exercises in this book with more detailed and useful information.

Foreword

I first met Paul Tran when I was busy creating another challenge in my life. I needed to take a vision from one man's mind, understand what the vision looked like, how it was going to work and comprehend the scale of his idea. My challenge was I was missing one very important ingredient, a tool that would create a picture with all the moving parts.

Research led me to discover a great tool, SOLIDWORKS. It claimed to allow one to make 3D components, in picture quality, on a computer, add in all moving parts, assemble it and make it run, all before money was spent on bending steel and buying parts that may not fit together. I needed to design and build a product with thousands of parts, make them all fit and work in harmony with millimeters tolerance. The possible cost implications of failed experimentation were daunting.

To my good fortune, one company's marketing strategy of selling a product without an instruction manual and requiring one to attend an instructional class to get it, led me to meet a communicator who made it all seem so simple.

Paul Tran has worked with and taught SOLIDWORKS as his profession for over 30 years. Paul knows the SOLIDWORKS product and manipulates it like a fine musical instrument. I watched Paul explain the unexplainable to baffled students with great skill and clarity. He taught me how to navigate the intricacies of the product so that I could use it as a communication tool with skilled engineers. *He teaches the teachers*.

I hired Paul as a design engineering consultant to create the machinery equipment with thousands of parts for my company's product. Paul Tran's knowledge and teaching skill has added immeasurable value to my company. When I read through the pages of these manuals, I now have an "instant replay" of his communication skill with the clarity of having him looking over my shoulder - *continuously*. We can now design, prove and build our product and know it will always work and not fail. Most important of all, Paul Tran helped me turn a blind man's vision into reality and a monument to his dream.

Thanks Paul.

These books will make dreams come true and help visionaries change the world.

Peter J. Douglas

CEO, Cake Energy, LLC

Images courtesy of C.A.K.E. Energy Corp., designed by Paul Tran

Author's Note

SOLIDWORKS 2019 Basic Tools and Advanced Techniques is comprised of lessons and exercises based on the author's extensive knowledge on this software. Paul has over 30 years of experience in the fields of mechanical and manufacturing engineering. He spent 2/3 of those years teaching and supporting the SOLIDWORKS software and its add-ins.

As an active Sr. SOLIDWORKS instructor and design engineer, Paul has worked and consulted with hundreds of reputable companies including IBM, Intel, NASA, US- Navy, Boeing, Disneyland, Medtronic, Edwards Lifesciences, Terumo, Kingston and many more. Today, he has trained more than 9000 engineering professionals, and given guidance to nearly half of the number of Certified SOLIDWORKS Professionals and Certified SOLIDWORKS Expert (CSWP & CSWE) in the state of California.

Every lesson and exercise in this book was created based on real world projects. Each of these projects has been broken down and developed into easy and comprehendible steps for the reader. Learn the fundamentals of SOLIDWORKS at your own pace, as you progress from simple to more complex design challenges. Furthermore, at the end of every chapter, there are self test questionnaires to ensure that the reader has gained sufficient knowledge from each section before moving on to more advanced lessons.

Paul believes that the most effective way to learn the "world's most sophisticated software" is to learn it inside and out, create everything from the beginning, and take it step by step. This is what the **SOLIDWORKS 2019 Basic Tools & Advanced Techniques** manuals are all about.

About the Training Files

The files for this textbook are available for download on the publisher's website at **http://www.sdcpublications.com/downloads/978-1-63057-224-2**.
They are organized by the chapter numbers and the file names that are normally mentioned at the beginning of each chapter or exercise. In the Built Parts folder you will also find copies of the parts, assemblies and drawings that were created for cross references or reviewing purposes.

It would be best to make a copy of the content to your local hard drive and work from these documents; you can always go back to the original training files location at anytime in the future, if needed.

Who this book is for

This book is for the mid-level user, who is already familiar with the SOLIDWORKS program. It is also a great resource for the more CAD literate individuals who want to expand their knowledge of the different features that SOLIDWORKS 2019 has to offer.

The organization of the book

The chapters in this book are organized in the logical order in which you would learn the SOLIDWORKS 2019 program. Each chapter will guide you through some different tasks, from navigating through the user interface, to exploring the toolbars, from some simple 3D modeling and move on to more complex tasks that are common to all SOLIDWORKS releases. There is also a self-test questionnaire at the end of each chapter to ensure that you have gained sufficient knowledge before moving on to the next chapter.

The conventions in this book

This book uses the following conventions to describe the actions you perform when using the keyboard and mouse to work in SOLIDWORKS 2019:

Click: means to press and release the mouse button. A click of a mouse button is used to select a command or an item on the screen.

Double Click: means to quickly press and release the left mouse button twice. A double mouse click is used to open a program or show the dimensions of a feature.

Right Click: means to press and release the right mouse button. A right mouse click is used to display a list of commands, a list of shortcuts that is related to the selected item.

Click and Drag: means to position the mouse cursor over an item on the screen and then press and hold down the left mouse button; still holding down the left button, move the mouse to the new destination and release the mouse button. Drag and drop makes it easy to move things around within a SOLIDWORKS document.

Bolded words: indicates the action items that you need to perform.

Italic words: Side notes and tips that give you additional information, or to explain special conditions that may occur during the course of the task.

Numbered Steps: indicates that you should follow these steps in order to successfully perform the task.

Icons: indicates the buttons or commands that you need to press.

SOLIDWORKS 2019

SOLIDWORKS 2019 is program suite, or a collection of engineering programs that can help you design better products faster. SOLIDWORKS 2019 contains different combinations of programs; some of the programs used in this book may not be available in your suites.

Start and exit SOLIDWORKS

SOLIDWORKS allows you to start its program in several ways. You can either double click on its shortcut icon on the desktop, or go to the Start menu and select the following: All Programs / SOLIDWORKS 2019 / SOLIDWORKS, or drag a SOLIDWORKS document and drop it on the SOLIDWORKS shortcut icon.

Before exiting SOLIDWORKS, be sure to save any open documents, and then click File/Exit; you can also click the X button on the top right of your screen to exit the program.

Using the Toolbars

You can use toolbars to select commands in SOLIDWORKS rather than using the drop-down menus. Using the toolbars is normally faster. The toolbars come with commonly used commands in SOLIDWORKS, but they can be customized to help you work more efficiently.

To access the toolbars, either right click in an empty spot on the top right of your screen or select View / Toolbars.

To customize the toolbars, select Tools / Customize. When the dialog pops up, click on the Commands tab, select a Category, and then drag an icon out of the dialog box and drop it on a toolbar that you want to customize. To remove an icon from a toolbar, drag an icon out of the toolbar and drop it into the dialog box.

Using the task pane

The task pane is normally kept on the right side of your screen. It displays various options like SOLIDWORKS resources, Design library, File explorer, Search, View palette, Appearances and Scenes, Custom properties, Built-in libraries, Technical alerts and news, etc.

The task pane provides quick access to any of the mentioned items by offering the drag and drop function to all of its contents. You can see a large preview of a SOLIDWORKS document before opening it. New documents can be saved in the task pane at anytime, and existing documents can also be edited and re-saved. The task pane can be resized, close or move to a different location on your screen if needed.

Become a CSWP – Certified SOLIDWORKS Professional

Demonstrate and validate your competency and knowledge in SOLIDWORKS software by achieving SOLIDWORKS certifications.

A Certified SOLIDWORKS Professional is an individual who has successfully passed our advanced skills examination. Each CSWP has proven their ability to design and analyze parametric parts and moveable assemblies using a variety of complex features in SOLIDWORKS software.

Standing out amongst SOLIDWORKS users from around the world can be challenging. Earning a SOLIDWORKS Certification can help you get a job, keep a job, or possibly move up in your current job. Be a part of our growing community of certified users. The certification center is your place to validate your certificate. This manual includes the CSWP-Exam preparation materials to help get you started.

For more information, log on to https://solidworks.virtualtester.com/#home_button

Table of Contents

SOLIDWORKS 2019

Advanced Modeling Topics

Chapter 3 **Advanced Modeling** **3-1**

Chapter 4 **Sweep with Composite Curves** **4-1**

Sheet Metal Topics

Mold Tools Design Topics

Glossary, Index, and SOLIDWORKS 2019 Quick-Guides
Quick Reference Guide to SOLIDWORKS 2019 Command Icons and Toolbars.

Introduction

SOLIDWORKS User Interface

The SOLIDWORKS 2019 User Interface

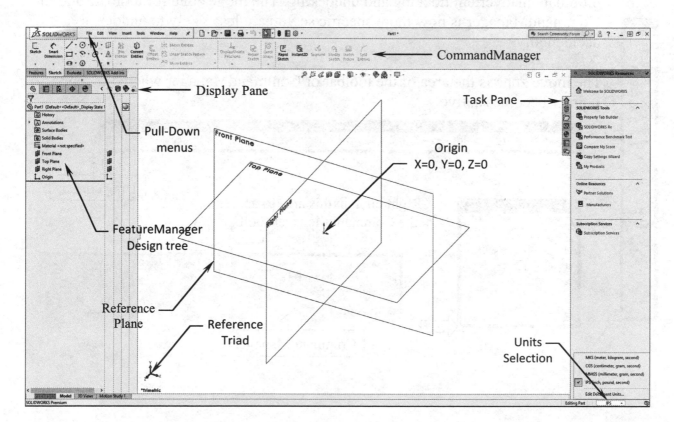

The 3 reference planes:
The Front, Top and the Right plane are 90 apart.
They share the same center point called the Origin.

Origin

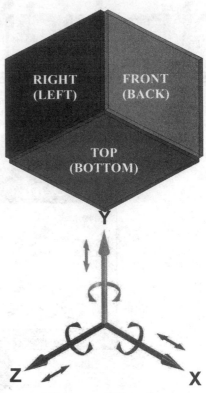

6 Degrees of Freedom

Docking Toolbars:

- The docking behavior of toolbars and the CommandManager has changed to eliminate inadvertent docking and undocking. The move zone for toolbars and the CommandManager is now more specific so you are less likely to undock accidentally.

- The move zone is the area of the toolbar or CommandManager where the move pointer becomes active.

Drag corner

Right click in this area to access the CommandManager toolbar

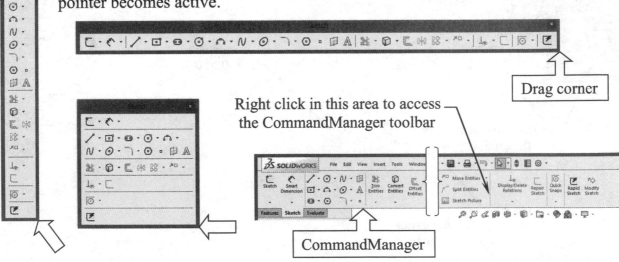

CommandManager

- If the CommandManager is not used, toolbars can be docked or left floating.

- Toolbars can be toggled off or on by activating or de-activating their check boxes:

- Select **Tools / Customize / Toolbars** tab.

- The icons in the toolbars can be enlarged when its check box is selected.

The View ports: You can view or work with a SOLIDWORKS model or an assembly using one, two, or four view ports.

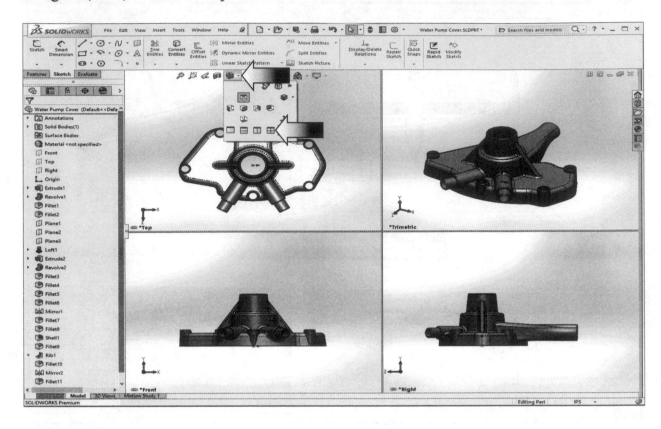

- Some of the **System Feedback symbols** (Inference pointers):

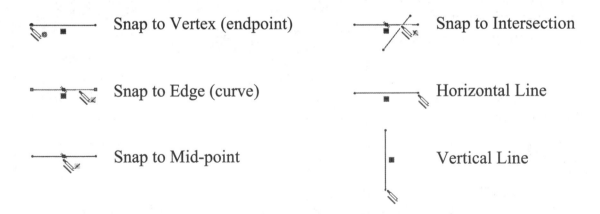

Snap to Vertex (endpoint)   Snap to Intersection

Snap to Edge (curve)   Horizontal Line

Snap to Mid-point   Vertical Line

The Status Bar: (View / Status Bar)

Displays the status of the sketch entity using different colors to indicate:

Green = Selected   **Blue** = Under defined
Black = Fully defined   **Red** = Over defined

<u>2D Sketch examples:</u>

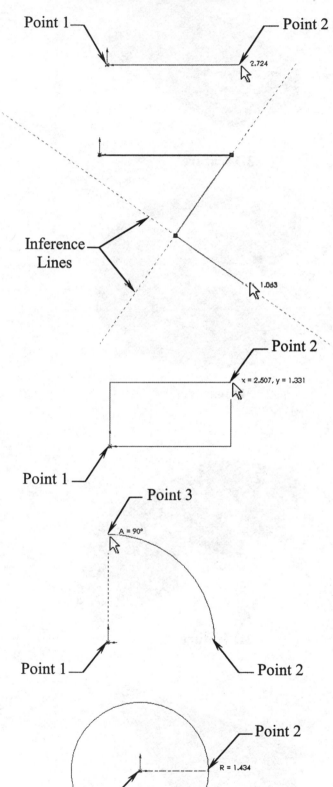

Click-Drag-Release: Single entity.

(Click Point 1, hold the mouse button, drag to point 2 and release.)

Click-Release: Continuous multiple entities.

(The Inference Lines appear when the sketch entities are Parallel, Perpendicular or Tangent with each other.)

Click-Drag-Release: Single Rectangle

(Click point 1, hold the mouse button, drag to Point 2 and release.)

Click-Drag-Release: Single Centerpoint Arc

(Click point 1, hold the mouse button and drag to Point 2, release; then drag to Point 3 and release.)

Click-Drag-Release: Single Circle

(Click point 1 [center of circle], hold the mouse button, drag to Point 2 [Radius] and release.)

3D Feature examples:

2D sketch → **Extrude** → **3D feature**

2D sketch → **Revolve** → **3D feature**

2D sketch → **Sweep** → **3D feature**

2D sketch → **Loft** → **3D feature**

Box-Select: Use the Select Pointer 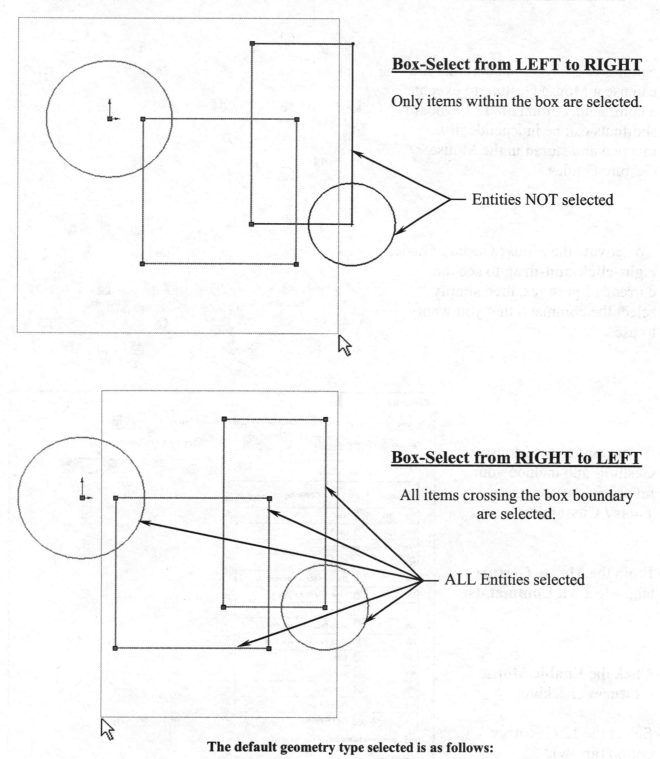 to drag a selection box around items.

Box-Select from LEFT to RIGHT

Only items within the box are selected.

Entities NOT selected

Box-Select from RIGHT to LEFT

All items crossing the box boundary are selected.

ALL Entities selected

The default geometry type selected is as follows:

* Part documents – edges * Assembly documents – components * Drawing documents - sketch entities, dims & annotations. * To select multiple entities, hold down **Ctrl** while selecting after the first selection.

The <u>Mouse Gestures</u> for Parts, Sketches, Assemblies and Drawings

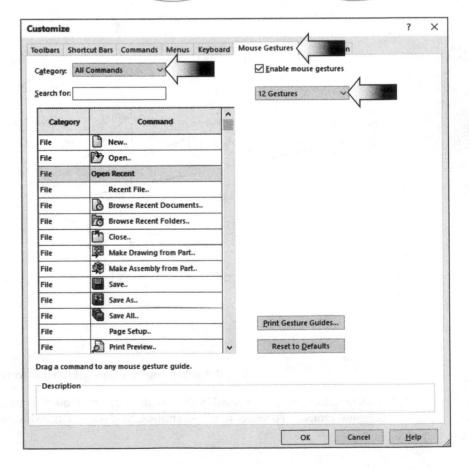

- Similar to a keyboard shortcut, you can use a Mouse Gesture to execute a command. A total of 12 keyboard shortcuts can be independently mapped and stored in the Mouse Gesture Guides.

- To activate the Mouse Gesture Guide, **right-click-and-drag** to see the current 12 gestures, then simply select the command that you want to use.

- To customize the Mouse Gestures and include your favorite shortcuts, go to: **Tools / Customize**.

- From the **Mouse Gestures** tab, select **All Commands**.

- Click the **Enable Mouse Gestures** checkbox.

- Select the **12 Gestures** option (arrow).

Customizing Mouse Gestures

To reassign a mouse gesture:

1. With a document open, click **Tools > Customize** and select the **Mouse Gestures** tab. The tab displays a list of tools and macros. If a mouse gesture is currently assigned to a tool, the icon for the gesture appears in the appropriate column for the command.

For example, by default, the right mouse gesture is assigned to the Right tool for parts and assemblies, so the icon for that gesture 🖰➔) appears in the Part and Assembly columns for that tool.

To filter the list of tools and macros, use the options at the top of the tab. By default, four mouse gesture directions are visible in the Mouse Gestures tab and available in the mouse gesture guide. Select 8 gestures to view and reassign commands for eight gesture directions.

2. Find the row for the tool or macro you want to assign to a mouse gesture, and click in the cell where that row intersects the appropriate column.

For example, to assign Make Drawing from Part 🖼 to the Part column, click in the cell where the Make-Drawing-from-Part row and the Part column intersect.

A list of either 4 or 12 gesture directions appears as shown, depending on whether you have the 4 gestures or 12 gestures option selected.

4 gestures   **12 gestures**

Some tools are not applicable to all columns, so the cell is unavailable and you cannot assign a mouse gesture. For example, you cannot assign a mouse gesture for Make Drawing from Part in the Assembly or Drawing columns.

3. Select the mouse gesture direction you want to assign from the list. The mouse gesture direction is reassigned to that tool and its icon appears in the cell.

4. Click OK.

Notes:

CHAPTER 1

Introduction To 3D Sketch

Introduction to 3D Sketch 🔳

SOLIDWORKS allows you to create 3D sketches. A 3D sketch consists of lines and arcs in series and splines. You can use a 3D sketch as a sweep path, as a guide curve for a loft or sweep, a centerline for a loft, or as one of the key entities in a piping system. Geometric relations can also be added to 3D Sketches.

Parameters

- **x**   **X Coordinate**
- **Y**   **Y Coordinate**
- **z**   **Z Coordinate**
- **Curvature** (Spline curvature at the frame point)
- **Tangency** (In the **XY** plane)
- **Tangency** (In the **XZ** plane)
- **Tangency** (In the **YZ** plane)

Space Handle

When working in a 3D sketch, a graphical assistant is provided to help you maintain your orientation while you sketch on several planes. This assistant is called a ***space handle***. The space handle appears when the first point of a line or spline is defined on a selected plane. Using the space handle you can select the axis along which you want to sketch.

Introduction to 3D Sketch

View Orientation Hot Keys:

Ctrl + 1 = Front View
Ctrl + 2 = Back View
Ctrl + 3 = Left View
Ctrl + 4 = Right View
Ctrl + 5 = Top View
Ctrl + 6 = Bottom View
Ctrl + 7 = Isometric View
Ctrl + 8 = Normal To
         Selection

Dimensioning Standards: **ANSI**
Units: **INCHES** – 3 Decimals

Tools Needed:

🔲 3D Sketch	🔲 2D Sketch	✏️ Sketch Line
◎ Circle	📏 Dimension	⊥ Add Geometric Relations
⌐ Sketch Fillet	Tab Tab Key	〰 Base/ Boss Sweep

1. Starting a new part file:

- Click **File / New.**

- Select the **Part** template and click **OK**.

2. Creating a 3D Sketch:

- Click or select **Insert / 3D Sketch**, and change to **Isometric view** .

- Select the Line tool and sketch the first line along the **X** axis.

—— Reference Axis
Indicator

- Sketch the second line along the **Y** axis as shown.

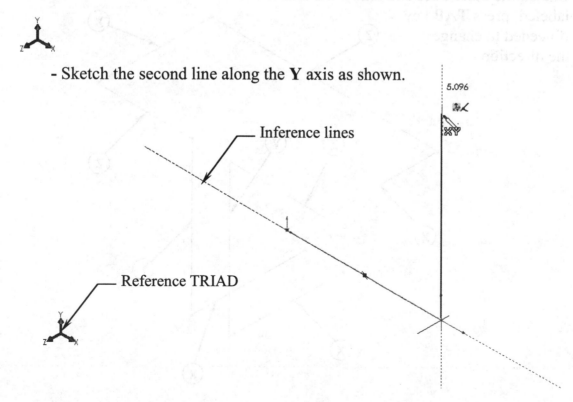

—— Inference lines

5.096

—— Reference TRIAD

3. Changing direction:

- By default your sketch is relative to the default coordinate system in the model.

- To switch to one of the other two default planes, press the **TAB** key and the reference origin of the current sketch plane is displayed on that plane.

Space Handle

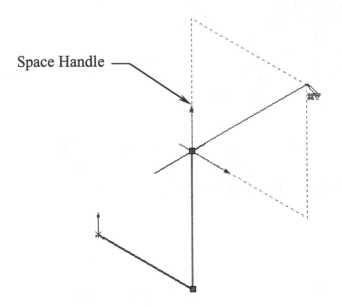

> 💡 **The TAB key**
>
> While sketching the lines, press the **TAB** key to switch to other planes/directions.

4. Completing the profile:

- Sketch the other lines and follow the axis as labeled; press **TAB** key if needed to change the direction.

5. Adding dimensions:

- Click or select **Tools / Dimensions / Smart Dimension**.

- Click on the first line and add a dimension of **3.00in**.

- There is not a general sequence
  to follow when adding dimensions,
  so for this lesson, add the dimensions
  in the same order you sketched the lines.

See
Note*

*_**Note:**_ To make the dimensions parallel to the lines
as shown, select the _line_ and an _endpoint_
instead of selecting just the endpoints.

- Continue adding the dimensions to
  fully define the 3D sketch as shown.

- Re-arrange the dimensions so
  they are easy to read, which will
  make editing a little easier later on.

6. Adding the Sketch Fillets:

- Click or select **Tools / Sketch Tools / Fillet**.

- Add **.500"** fillets to <u>all</u> the intersections as indicated.

- Enable the **Keep Constrained Corner** check box (Maintains the virtual intersection point if the vertex has dimensions or relations).

- Click **OK** when finished.

- **Exit** the 3D Sketch or press **Control + Q**.

🔆 Geometric Relations

Geometric Relations such as Along X, Y, Z and Equal can also be used to replace some of the duplicate dimensions.

7. Creating the Swept feature:

- The new Circular Profile sweep option allows you to create a solid rod or hollow tube along a path, edge, or curve directly on a model without having to sketch the circular profile. This enhancement is available for Swept Boss/Base, Swept Cut, and Swept Surface features.

- Click or select **Insert / Boss-Base / Sweep**.

- Select the **Circle Profile** option and enter **.250in** for the diameter of the profile ⊘.

- Select the **3D Sketch** for Sweep Path ↻ (3Dsketch1).

- Click **OK**.

8. Saving your work:

- Select **File / Save As**.

- Enter **3D Sketch** for the file name.

- Click **Save**.

Questions for Review

Introduction to 3D Sketch

1. When using 3D Sketch you do not have to pre-select a plane as you would in 2D Sketch.
   a. True
   b. False

2. The space handle appears only after the first point of a line is started.
   a. True
   b. False

3. To switch to other planes in 3D Sketch mode, press:
   a. Up Arrow
   b. Down Arrow
   c. TAB key
   d. CONTROL key

4. Dimensions cannot be used in 3D Sketch mode.
   a. True
   b. False

5. Geometric Relations cannot be used in 3D Sketch mode.
   a. True
   b. False

6. All sketch tools in 2D Sketch are also available in 3D Sketch.
   a. True
   b. False

7. When adding sketch fillets, the option Keep Constrained Corner will create a virtual intersection point but will not create a dimension.
   a. True
   b. False

8. 3D Sketch entities can be used as a path in a swept feature.
   a. True
   b. False

7. FALSE
8. TRUE
5. FALSE
6. FALSE
3. C
4. FALSE
1. TRUE
2. TRUE

Exercise: Sweep with 3D Sketch

1. Create the part shown using 3D Sketch.

2. Save your work as:
   Sweep w_3D Sketch_Exe.

Exercise: 3D Sketch & Planes

A 3D sketch normally consists of lines and arcs in series, and splines. You can use a 3D sketch as a sweep path, as a guide curve for a loft or sweep, a centerline for a loft, or as one of the key entities in a routing system.

The following exercise demonstrates how several planes can be used to help define the directions of 3D Sketch Entities.

1. Sketching the reference Pivot lines:

- Select the <u>Top</u> plane and open a new sketch .

- Sketch **2 Centerlines** and add Dimensions as shown.

2. Creating the 1st 45° Plane:

- Select **Insert/Reference Geometry/Planes** .

- Click the **At Angle** button and enter **45** for Angle (arrow).

- Select the **Top** plane and the **Vertical line** as noted.

Select the top plane and the vertical line…

- Click **OK**.

3. Creating the 2nd 45° Plane:

- Click the **Plane** command or select **Insert/Reference Geometry/Planes** .

- Click the **At Angle** option and enter **45** for Angle (arrow).

- Select the **Front** plane and the **Horizontal line** as noted.

Select the front plane
and the horizontal line

- Click **OK**.

Starting point
(At the endpoint
of the centerline).

1st line

1.534
1

4. Creating the 3D Sketch:

- Select the <u>Top</u> plane
  and click **Insert/3D Sketch** 3D .

- Sketch the 1st line along the **Z** direction as
  noted (or Vertical along the Top plane).

- Select the **Plane2** (45 deg.) from the Feature Manager tree and Sketch the 2nd line along the **Y** direction (watch the cursor feedback symbol).

> 💡 **Switching Planes**
>
> While sketching the lines, hold the **Control** key and click a plane to switch from one plane to another, or simply select them from the Feature tree each time.

- Sketch the rest of the lines on the planes as labeled.

- For clarity, hide all the planes (select **View / Hide-Show** and click off **Planes**). We will select the planes from the FeatureManager tree when needed.

Top View (Ctrl + 5)   Right View (Ctrl + 4)

- Add Dimensions 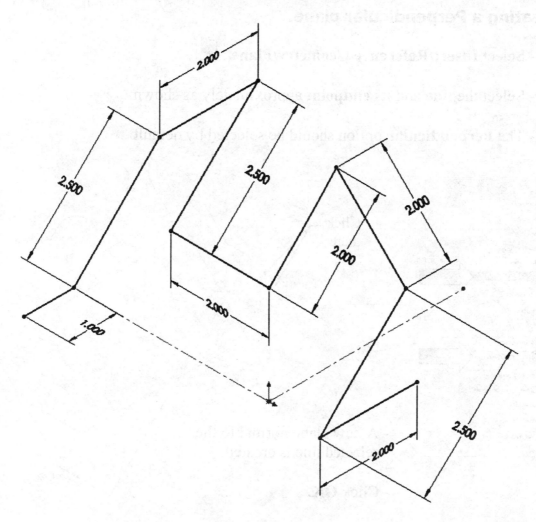 to fully define the sketch.

- Add **Sketch Fillets** of **.500 in.** to <u>all</u> corners.

- **Exit** the 3D Sketch or press **Ctrl+Q**.

5. Creating a Perpendicular plane:

- Select **Insert/Reference Geometry/Plane** .

- Select the **line** and its **endpoint** approximately as shown.

- The **Perpendicular** option should be selected by default.

Click here

- A new plane normal to the selected line is created.

- Click **OK**.

6. Sketching the Sweep Profile:

- Select the <u>new plane</u>
   (Plane3) and open a new sketch.

- Sketch **2 Circles**
   on the same center and
   add the dimensions as
   shown to fully define
   the sketch.

7. Sweeping the Profile along the 3D Path:

- Click or Select **Insert/Boss Base/Sweep**.

- Select the **Circles** as the Sweep Profile .

- Select the **3D Sketch** as the Sweep Path .

- Click **OK**.

- The resulting Swept feature.

8. Hiding the Planes:

- Select **View / Hide-Show / Planes**.

- The planes are temporarily put away from the scene.

9. Saving your work:

- Select **File / Save As**.

- Enter **3D Sketch_Planes** for the name of the file.

- Click **Save**.

Exercise: 3D Sketch & Composite Curve

A 3D sketch normally consists of lines and arcs in series and Splines. You can use a 3D sketch as a sweep path, as a guide curve for a loft or sweep, a centerline for a loft, or as one of the key entities in a routing system.

The following exercise demonstrates how several 3D Sketches can be created, combined into 1 continuous Composite Curve, and used as a Sweep Path.

1. Creating a 2D sketch:

- Select <u>Top</u> plane and sketch a **1.00in** diameter **Circle** and 2 Centerlines.

2. Creating a Helix:

- Select **Insert/Curve/ Helix-Spiral**.

- Pitch: **.250 in**.

- Revolution: **10**.

- Starting Angle: **0 deg**.

- Click **OK**.

3. Creating the 1st 3D sketch:

- Select **Insert/3D Sketch** 3D

- Select the **Line** command
  and sketch the 1st line
  along the X direction.

On-Plane relation
(End point & Right plane)

- Add other lines in the directions as labeled.

- Add Dimensions to fully define the sketch.

Sketch Fillet

✓ ✕ ✗

Message
Select a sketch vertex or entities to fillet.

Entities to Fillet
Fillet<1>
Fillet<2>
Fillet<3>
Fillet<4>
Fillet<5>
Fillet<6>
Fillet<7>

Fillet Parameters
0.250in
☑ Keep constrained corners
☐ Dimension each fillet

- Add Sketch Fillets
  of **.250 in**. to all corners.

- **Exit** the 3D Sketch 3D or press **Ctrl + Q**.

4. Creating the 2nd 3D sketch:

- Select **Insert/3D Sketch** [3D] .

- Select the **Line** command [/] and sketch the 1st line along the X direction.

- Sketch the rest of the lines following their direction shown below.

On-Plane relation
(End point & Right plane)

- Add Dimensions [↙] to fully define the sketch.

- Add **Sketch Fillets** [◥] of **.250 in.** to all corners.

- **Exit** the 3D Sketch [3D] or press **Ctrl+Q**.

1-19

5. Combining the curves:

- Select the **Composite Curve** command below the Curves button or select:
  Insert / Curve / Composite.

- Select the 3 Sketches either from the Feature Manager tree or directly from the graphics area.

- Click **OK**.

- The sketches are now
  combined into 1 continuous curve.
  We will use it as the sweep path in the few steps.

6. Creating a Sweep using Circular Profile:

- Select **Insert/Boss Base/ Sweep** .

- Select the **Circle Profile** option (arrow).

- Enter **.165 in** for the diameter of the sweep profile .

- Select the **Composite Curve** as the Sweep Path .

- Click **OK**.

7. Saving your work:

- Click **File/Save As**.

- Enter **3D Sketch_ Composite Curve** for the name of the file.

- Click **Save**.

CHAPTER 2

Plane Creation

Plane Creation
Advanced Topics

- In SOLIDWORKS, planes are not only used to sketch geometry but are also used to create section views of a model or an assembly. Planes are also used as the end conditions to extrude features and as neutral planes to define the draft angles, etc.

- There are several options to create planes:

 Parallel Plane.   At Angle Plane.

 Perpendicular Plane.   Offset Distance Plane.

 Coincident Plane.   Mid Plane.

 Project Plane.

- Each plane requires slightly different types of references; some of them may require only one and some others may require two or three.

- This chapter discusses how planes are created using sketch geometry and other features that are available in the model, as references.

Plane Creation

Dimensioning Standards: **ANSI**

Units: **INCHES** – 3 Decimals

Tools Needed:

Insert Sketch	Rectangle	Circle
Planes	Add Geometric Relations	Dimension
Sketch Mirror	Offset Entities	Boss/Base Revolve
Circular Pattern	Extruded Cut	Fillet/Round

1. Starting with a new Part document:

- Select **File / New / Part** and click **OK**.

- Select the <u>Front</u> plane from the FeatureManager tree.

- Click ⬚ or select **Insert / Sketch**.

- Sketch the profile below and add dimensions as shown.
  (It might be easier to sketch a circle, instead of a center-point arc, add the 2 centerlines, and then trim off the bottom half of the circle).

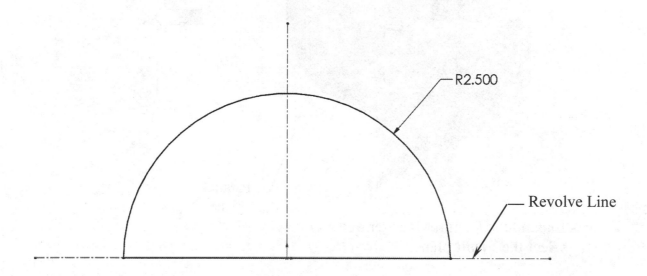

R2.500

Revolve Line

2. Revolving the Base:

- Click 🌀 or select **Insert / Boss Base / Revolve**.

- Set Revolve Type to: **Blind**.

- Set Revolve Angle to: **360 deg**.

- Click **OK**.

- *Note: Drag the Direction arrow to see the preview of the rotate angle.*

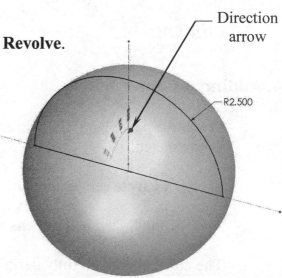

Direction arrow

R2.500

3. Creating a Tangent plane: (Requires a cylindrical face and a parallel plane).

- Click 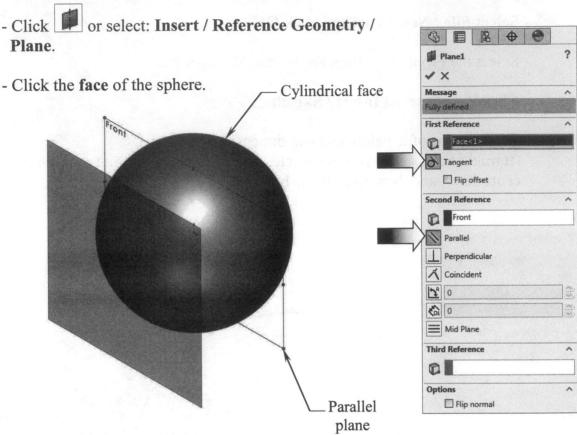 or select: **Insert / Reference Geometry / Plane**.

- Click the **face** of the sphere.

Cylindrical face

Parallel plane

- Expand the FeatureManager tree and select the **Front** plane.

- The **Tangent** option is selected automatically.

- Click the **Parallel** option in the Second Reference section.

- Click **OK**.

4. Adding a Center hole:

- Select the <u>new plane</u> (Plane1) and open a new sketch.

- Sketch a **Circle** centered on the origin.

- Add a **1.000"** diameter dimension.

- The circle should be fully defined at this point.

- Click or select:
Insert / Cut / Extrude.

- Select **Through All** for
Direction 1.

- Click **OK**.

5. Creating a flat surface:

- This step will demonstrate the use of geometric relations to fully define the sketch
without using dimensions.

- Select the <u>Right</u> plane from the FeatureManager tree.

- Click or select **Insert / Sketch** and switch to the right view (Ctrl+4).

- Sketch a **Line** and add the Relations as shown.

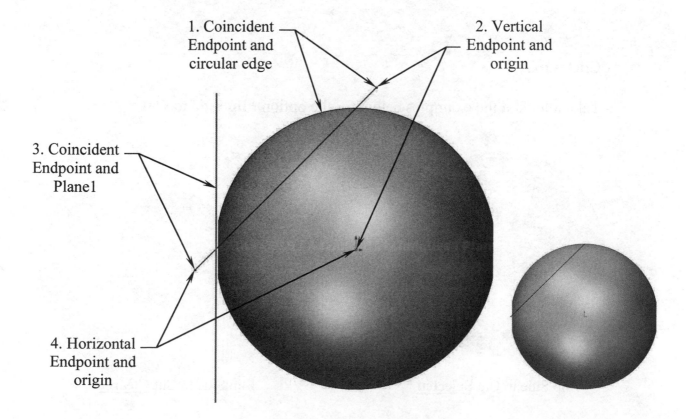

1. Coincident Endpoint and circular edge

2. Vertical Endpoint and origin

3. Coincident Endpoint and Plane1

4. Horizontal Endpoint and origin

6. Extruding a Cut:

- Click 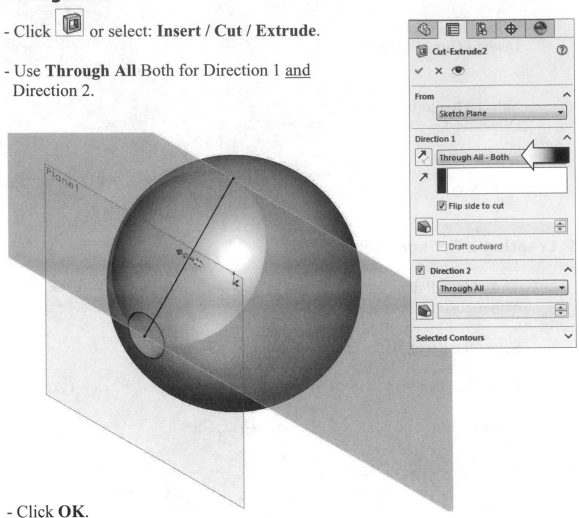 or select: **Insert / Cut / Extrude**.

- Use **Through All** Both for Direction 1 <u>and</u>
  Direction 2.

- Click **OK**.

- Take a look at the examples below for the option **Flip Side to Cut**.

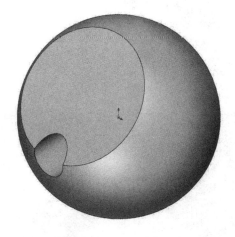

Flip Side to Cut <u>**Selected**</u>

Flip Side to Cut <u>**Cleared**</u>

7. Creating an At-Angle plane: (Requires a Reference Plane, a Reference Axis, and an Angular Dimension).

    - Select the **Front** plane from FeatureManager tree.

    - Click [icon] or select **Insert / Reference Geometry / Plane**.

    - Select the **horizontal centerline** as reference axis.

    - Select the **At Angle** option.

    - Enter **15 deg**. in the dialog box and click **Flip Offset**.

    - Click **OK**.

> 💡 **Show Sketch** 👓
>
> Right click on *Sketch1* (on the FeatureManager tree, below the **Revolve1**) and select **Show**.

Reference Axis —

8. Creating a Ø.750 hole:

- Select the <u>new plane</u> (Plane2).

- Click [icon] or select **Insert / Sketch**.

- Sketch a **Circle** [icon] and add the dimensions and a vertical relation as shown.

.950

Plane2

Ø.750

Add a ref. centerline to use later in step 10.

Vertical —

- Click 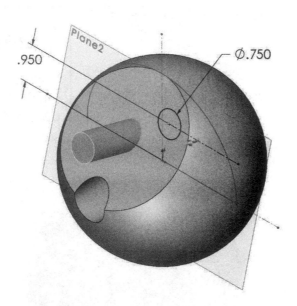 or select **Insert / Cut / Extrude**.

- Direction 1: **Through All Both**.

- Click **OK**.

9. Showing the Sketches:

- On the FeatureManager tree expand the Cut-Extrude1 (click the + symbol), right click on **Sketch2**, and select **Show**.

- Expand the Cut-Extrude2 (click the + symbol), right-click on **Sketch3**, and select **Show** ; also **Hide** the Sketch1.

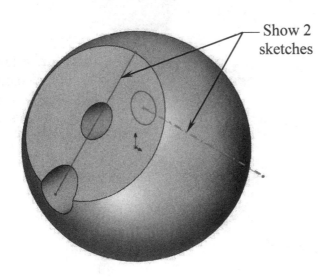

Show 2 sketches

10. Creating a Coincident plane: (Requires a Reference Line and a Sketch Point or a Vertex).

- Click 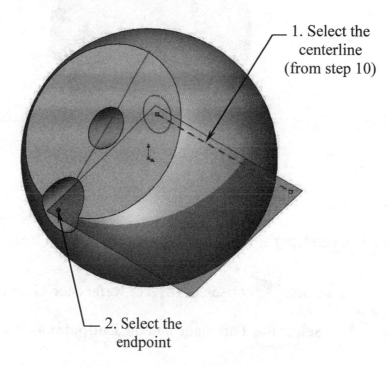 or select **Insert / Reference Geometry / Plane**.

- Select the **Centerline** and the **Endpoint** as indicated.

- The **Coincident** option should be selected automatically.

- Click **OK**.

1. Select the centerline (from step 10)

2. Select the endpoint

11. Creating the Ø.500 holes:

- Select the <u>new plane</u> (Plane3).

- Click ⊞ or select **Insert / Sketch**.

- Sketch a **Circle** ⊙ and Mirror it 🔀 (Use either Dynamic Mirror or Mirror Entity to mirror the circles).

- Add Dimensions 📏 as shown to fully define the sketch.

Mirror Centerline

1.000

1.000

Ø.500

- Click or select **Insert / Cut / Extrude**.

- Direction 1: **Through All Both**.

- Click **OK**.

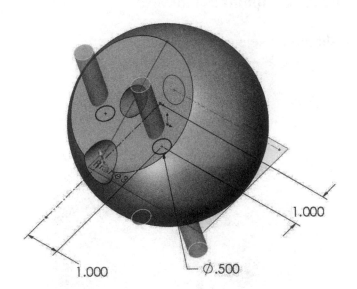

12. Creating a Parallel plane: (Requires a Reference Plane and Reference Point).

- Click or select **Insert / Reference Geometry / Plane**.

- Select the **Top** plane and the **Endpoint** as indicated.

- Based on the selection, the system selects the **Parallel** and **Coincident** options.

- Click **OK**.

1. Select the TOP plane

2. Select the Endpoint

13. Creating the Ø2.500 Recess:

- Select the <u>new Plane</u> (Plane4) and insert a new sketch .

- Sketch a **Circle** and add Dimension .

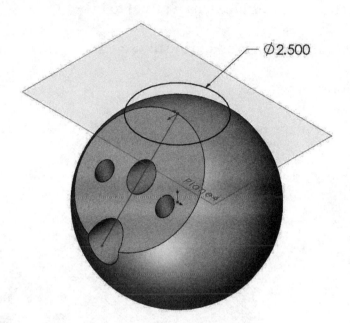

- Click or select **Insert / Cut / Extrude**.

- End Condition: **Blind**.

- Extrude Depth: **.625** in.

- Click **OK**.

- **Hide** the Sketch2, Sketch3 and all planes.

14. Creating an Offset-Distance plane: (Requires a Reference Plane and a Distance dimension).

- Click or select **Insert / Reference Geometry / Plane**.

- Select **Plane3** (from the FeatureManager tree) to offset from.

- The **Offset Distance** option is automatically selected.

- Enter **3.375** for offset value.

- Make sure the new plane is placed below the Plane3 (click Flip if needed).

- Click **OK**.

15. Creating the Bore holes:

- Select the <u>new plane</u> (Plane5) and insert a new sketch .

- Select the **circular edge** of the hole and press **Offset-Entities** .

- Enter **.100 in**. for Offset Distance (Only one offset can be done at a time, since the 2 circles are not connecting to each other).

- Click 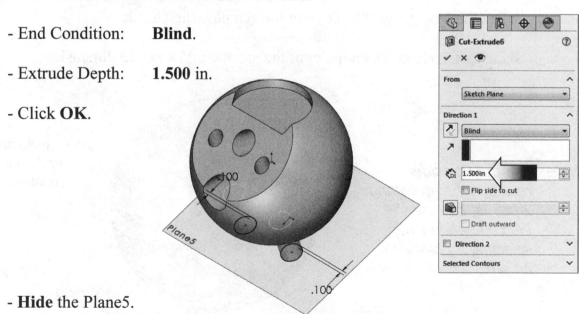 or select **Insert / Cut Extrude**.

- End Condition: **Blind**.

- Extrude Depth: **1.500** in.

- Click **OK**.

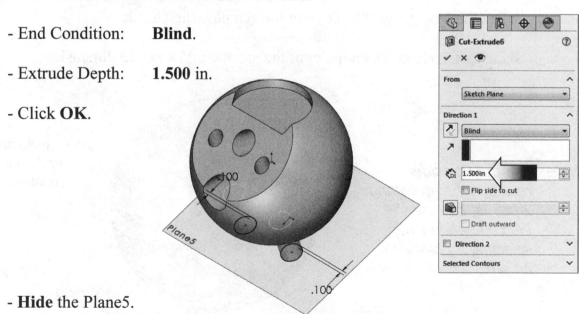

- **Hide** the Plane5.

16. Creating a Perpendicular plane: (Requires a Reference Line or Curve & a Point).

- Click or select **Insert / Reference Geometry / Plane**.

- **Show** the **Sketch1** and select the **Arc** and the **Endpoint** as noted.

- The **Perpendicular** and **Coincident** options should be selected automatically.

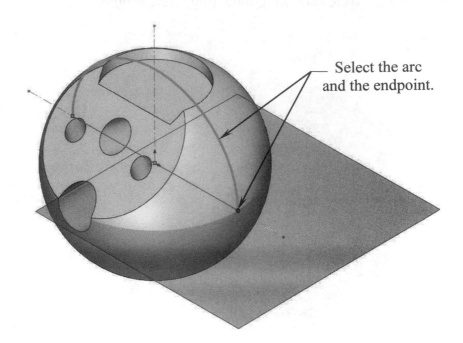

Select the arc and the endpoint.

- Click **OK**.

17. Creating the side-grips:

- Select the <u>new plane</u> (Plane6) and insert a new sketch ⊞ .

- Sketch a **Circle** at the endpoint of the arc and add a **Ø.625** dimension.

Note: *Use the Coincident relation when selecting 2 points, but use the Pierce relation when selecting a point and an arc.*

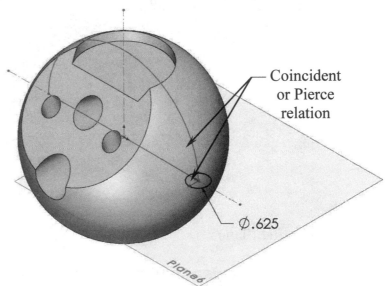

Coincident or Pierce relation

Ø.625

- Click ⊞ or select **Insert / Cut Extrude**.

- Direction 1: **Through All Both**.

- Click **OK**.

- **Hide** the Sketch1 and the Plane6.

18. Creating a Circular Pattern of the Grips:

- Click or select **Insert / Pattern Mirror / Circular Pattern**.

- Click **View / Temporary Axis** and select the center **axis** as indicated.

- Equal Spacing: **Enabled**.

- Total Angle: **360 deg**.

- Number of instances: **12**.

- Select the Cut-Extrude6 as Feature To Pattern.

- Click **OK**.

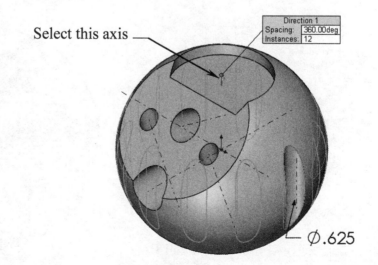

Select this axis

19. Adding another recess:

- Select the <u>upper face</u> of the recess and insert a new sketch .

- Sketch a **Circle** centered on the Origin.

- Add a **1.750 diameter** dimension.

Ø1.750

Sketch face

- Click or select **Insert / Cut Extrude**.

- End Condition: **Blind**.

- Extrude Depth: **.175** in.

- Click **OK**.

20. Creating a Mid-Plane: (Requires 2 parallel planes or 2 planar faces).

- Click or select:
Insert / Reference Geometry / Plane.

- Expand the Feature-Manager and select the **Plane3** and the **Plane5** to use as the first and second references.

- The **Mid-Plane** option is selected automatically; if not, click the mid plane button (arrow).

- A preview of the new plane appears in the middle of the two selected planes.

- Click **OK**.

21. Creating a rectangular pocket:

- Select the <u>new plane</u> (Plane7) and insert a new sketch .

- Sketch a **Center Rectangle** that is centered on the origin.

- Add the width and height dimensions to fully define the sketch.

.675

1.000

- Click or select **Insert / Cut Extrude**.

- End Condition: **Through All**.

- Reverse Direction: **Enabled** (arrow).

- Draft: **1deg. Outward** (arrow).

- Click **OK**.

22. Adding fillets to the pocket:

- Click or select **Insert Features / Fillet-Round**.

- Use the default **Constant Radius** option.

- Enter **.093in.** for radius.

- Select the **4 edges** of the pocket as noted.

Select
4 edges

- Enable the Full Preview checkbox.

- Click **OK**.

23. Adding Fillets to all edges:

- Lasso or Box-Select around the part,
  (or press Ctrl+A), to select all of its edges.

- Click or select **Insert Features / Fillet-Round**.

- Enter **.040** for Radius.

- Tangent Propagation: **Enabled**.

- Click **OK**.

- Hide the planes
  before saving the
  part. **(View / Hide-
  Show / Planes)**.

24. Saving your work:

- Click **File / Save As / Planes Creation / Save**.

Questions for Review

Plane Creation

1. Planes can be used to section a part or an assembly.
   a. True
   b. False

2. A sketch can be extruded to a plane as the end condition by using the Up-To-Surface option.
   a. True
   b. False

3. Which one of the options below is not a valid command?
   a. Parallel plane at Point.
   b. Offset plane at Distance.
   c. Perpendicular to another plane at Angle.
   d. Normal to Curve.

4. To create a plane at Angle, you will need:
   a. The Angle and a Reference plane.
   b. The Angle and a pivot Line.
   c. The Angle, a pivot Line, and a Reference plane.

5. To create a plane through Lines/Points, you will need at least:
   a. One line and a point
   b. Two lines and a point
   c. Two lines and Two points

6. To create a Parallel Plane At Point, you will need a reference plane and a point.
   a. True
   b. False

7. When creating a Plane Normal To Curve, you can select:
   a. A linear model edge.
   b. A straight line.
   c. A 2D or 3D curve.
   d. All of the above.

7. D
5. A   6. TRUE
3. C   4. C
1. TRUE   2. TRUE

Section with Front plane

Section with Right plane

Section with Top plane

Isometric View

Exercise: Creating New Planes

1. Create a reference sketch:

- Create the sketch shown on the **Front** plane.

2. Create 3 new planes:

- Use the references as indicated.

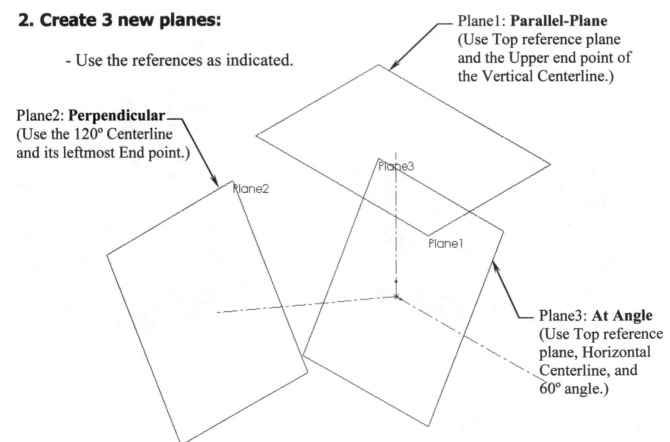

Plane1: **Parallel-Plane**
(Use Top reference plane and the Upper end point of the Vertical Centerline.)

Plane2: **Perpendicular**
(Use the 120° Centerline and its leftmost End point.)

Plane3: **At Angle**
(Use Top reference plane, Horizontal Centerline, and 60° angle.)

3. Open a new sketch on the Plane1:

- Sketch a circle centered on the end point of the vertical centerline.

- Add a the **2.250in** diameter dimension.

4. Extrude a boss:

- Type: **Blind**.

- Depth: **8.000in**

- **Reverse** direction enabled.

- Click **OK**.

5. Open a new sketch on the Plane2:

- Sketch a circle centered on the left end of the 120deg. centerline.

- Add a the **1.750in** diameter dimension.

6. Extrude a boss:

- Type: **Up To Surface**.

- Select the **outer face** of the 1st cylinder.

Select face

- Click **OK**.

7. Open a new sketch on the plane3:

- Sketch a **Corner Rectangle** as shown.

- Add the height and the width dimensions.

8. Extrude a boss:

- Type: **Blind**.

- Depth: **1.000in**.

- Click **OK**.

9. Add the .250in fillets:

- Add a fillet of **.250in** to the **4 edges** as noted.

Select 4 edges

- Click **OK**.

10. Add the .125in fillets:

- Add a fillet of **.125in** to the **2 edges** as indicated below.

Select 2 edges

- Click **OK**.

11. Shell the model:

- Shell the model using a thickness of **.080in**.

- Select the **4 faces** to remove as indicated.

Select 4 faces

- Click **OK**.

12. Save your work:

- Click **File / Save As**.

- Enter **Plane_Exe.sldprt** for the file name.

- Click **Save**.

- Close all documents.

CHAPTER 3

Advanced Modeling

Advanced Modeling
5/8" Spanner

- The draft option is omitted in this lesson to help focus on other areas.

- The arc conditions Min / Max are options that help place the dimensions
  on the tangents of arcs or circles. Once a dimension is created, the arc conditions
  can be changed by right clicking on the dimension and selecting the Leaders tab.
  Only two conditions can be specified at a time.
  Either Center/Center, Min/Max, Max/Max,
  or Min/Min, etc.

- Adding text on the model is another
  unique feature in SOLIDWORKS. This
  option allows the letters in the sketch to be
  extruded as an emboss or a cut, similar to
  other extruded features.

- All letters in the same sketch are treated as one entity; they
  will be extruded at the same time and will have the same
  extrude depth. However, the option Dissolve-Sketch-
  Text is used to convert the sketch text into sketch entities
  so that the shape and size of each letter can be modified
  individually.
  To use this option simply right click the text and select:
  Dissolve Sketch Text.

- This chapter and its exercise will guide you through some of the advanced
  modeling techniques as well as learning to use the Text tool to create the straight
  or curved extruded letters.

Advanced Modeling
5/8" Spanner

View Orientation Hot Keys:

Ctrl + 1 = Front View
Ctrl + 2 = Back View
Ctrl + 3 = Left View
Ctrl + 4 = Right View
Ctrl + 5 = Top View
Ctrl + 6 = Bottom View
Ctrl + 7 = Isometric View
Ctrl + 8 = Normal To
        Selection

Dimensioning Standards: **ANSI**

Units: **INCHES** – 3 Decimals

Tools Needed:

Insert Sketch	Line	3 Point Arc
Text	Add Geometric Relations	Dimension
Sketch Fillet	Polygon	Plane
Base/Boss Extrude	Extruded Cut	Fillet/Round

1. Opening the Spanner sketch document:

- From the Training Files folder, locate and <u>open</u> the document named: **Spanner Sketch**.

- <u>**Edit**</u> the **Sketch1**. This is the open end of the spanner.

Symmetry centerline

Maximum Arc Condition

Tangent Relations

Perpendicular with the 10° centerline

💡 **Arc Conditions**

The 1.375 dimension is measured from the outer tangents of the arcs. This is called Maximum Arc Conditions*

---

* There are two options to create the Max/Min Arc condition dimensions:

Option 1: Select the Smart Dimension tool, click on the 2 arcs, and place the dimension anywhere.
Select the Leaders tab from the tree (circled).
Select the Max options for both arcs (Circled).

Option 2: Hold down the SHIFT key and click on the 2 arcs; the dimension's leader lines will snap to the arc tangents automatically.

2. Extruding the base feature:

- Click or select **Insert / Boss-Base / Extrude**.

- End Condition: **Mid Plane**.

- Extrude Depth: **.250 in**.

- Click **OK**.

3. Creating the transition sketch:

- Select Top plane from the FeatureManager tree.

- Click or select **Insert / Sketch**.

- Sketch the profile below using Lines [/].

Note: only add the Sketch Fillets after the sketch is fully defined.

- Add dimensions [◢] or Relations [⊥] as needed.

Line tangent with Arc

Virtual Intersections*

6.250

1.000

.750

Convert 3 entities

.525   .950

Sketch fillets

R2.000

Trim Entities as needed

R4.000

4. Extruding the Transition feature:

- Click or select **Insert / Boss-Base / Extrude**.

- End Condition: **Mid Plane**

- Extrude Depth: **.175 in**.

- Click **OK**.

5. Adding the reference geometry:

- Select the <u>face</u> as indicated.

- Click ⌐ or select **Insert / Sketch**.

- Sketch a **Centerline** ✎ at the mid-point of the two vertical edges.

Sketch Face

Mid-point

- <u>Exit</u> the Sketch ⌐ or select **Insert / Sketch**.

6. Creating a new work plane: Plane at Angle

- Click or select **Insert / Reference Geometry / Plane**.

- For Reference Entities, select the **Sketch4** (centerline) and the **upper face** of the Transition.

- Enter **10 deg**. and click **Flip Offset** to place the plane on the bottom.

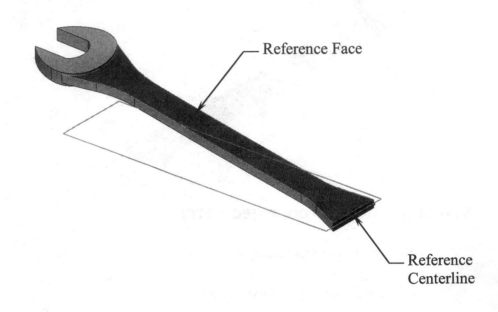

- Click **OK**.

- SOLIDWORKS creates a plane that starts from the reference face and pivots around the centerline.

- The preview of the new 10 degrees plane.

7. Creating the Closed-End sketch:

- Select the new **10° plane** from the FeatureManager tree.

- Click or select **Insert / Sketch**.

- Sketch a circle ⊙ and add dimensions ↖ as shown.

6.500

Ø1.000

Vertical

6.500

Ø1.000

8. Extruding the Closed-end feature:

- Click ⬓ or select **Insert / Boss-Base / Extrude**.

- **Direction 1:**   **Blind**.

- Extrude Depth: **.200 in**.

- **Direction 2:**   **Blind**.

- Extrude Depth: **.130 in**.

- Click **OK**.

6.500

Ø1.000

9. Adding a 12-Sided polygonal hole:

- Select the <u>face</u> indicated as sketch plane.

- Click 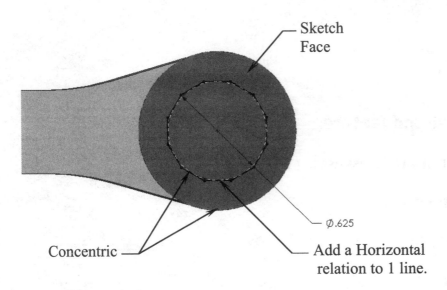 or select **Insert / Sketch**.

- Sketch a Polygon with **12 sides** (arrow).

- Add a **.625 Dia**. Dimension to the inside construction circle.

- Add a **Concentric** relation between the construction circle and the circular edge.

Sketch Face

Ø.625

Concentric

Add a Horizontal relation to 1 line.

10. Extruding a cut:

- Click or select **Insert / Cut / Extrude**.

- End Condition: **Through All**.

- Click **OK**.

Ø.625

11. Creating the Recess profile:

- Select the <u>face</u> indicated as sketch plane.

- Click [icon] or select **Insert / Sketch**.

- Sketch the profile shown below using the **Straight-Slot** command [icon].

Sketch face

- Add Dimensions [icon] and Relations [icon] to fully define the sketch.

12. Extruding the Recessed feature:

- Click [icon] or select **Insert / Cut / Extrude**.

- End Condition: **Blind**.

- Extrude Depth: **.030 in**.

- Click **OK**.

13. Mirroring the Recessed feature:

- Hold the **Control** key, select the <u>Top</u> reference plane and the Recessed feature from the FeatureManager tree.

- Click or select **Insert / Pattern Mirror** menu, then select **Mirror**.

- Click **OK**.

- Rotate the model to verify the results.

14. Adding the .030" fillets:

- Click or select **Insert / Features / Fillet/Round**.

- Enter **.030 in**. for Radius.

- Select the edges as shown for Edges to fillet.

- Tangent Propagation: **Enabled**.

- Click **OK**.

Select these edges
on both sides…

15. Adding the .050" fillets:

- Repeat step 14 and add a **.050"** fillet to the 4 edges shown below.

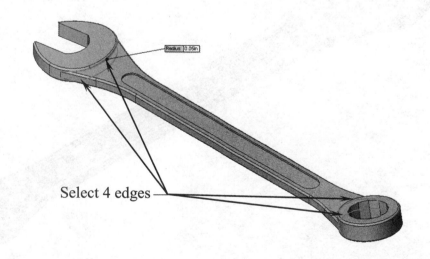

Select 4 edges

16. Adding the .015" fillets:

- Click 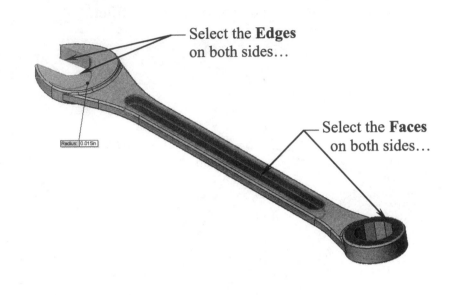 and add a **.015"** fillet to the edges and faces shown below.

Select the **Edges** on both sides…

Select the **Faces** on both sides…

- Click **OK**.

- Verify your fillets with the model shown below.

17. Adding text:

- Select the <u>face</u> indicated as sketch plane.

- Click or select **Insert / Sketch**.

- Click and type **5/8** in the text dialog box.

- Click **OK**.

- Add dimensions to position the text.

> ☼ **Positioning Text**
>
> Each set of sketch text comes with a Manipulator Point; dimensions can be added from this point to position the text.

Sketch Face

5/8

.075

1.325

Manipulator Point

- Clear Use document's font check box [ ] Use document's font .

- Change Width factor to **150%** [A] .

- Leave Spacing at **100%** [AB] .

- Font: **Century Gothic** [Font...] .

- Style: **Regular** - Points size: **14 pt.**

<u>NOTE:</u>
Use the Curves option when you want your sketch letters to wrap along a curve.

It will work better if the curve is created in the same sketch, as construction geometry.

18. Extruding the text:

- Click 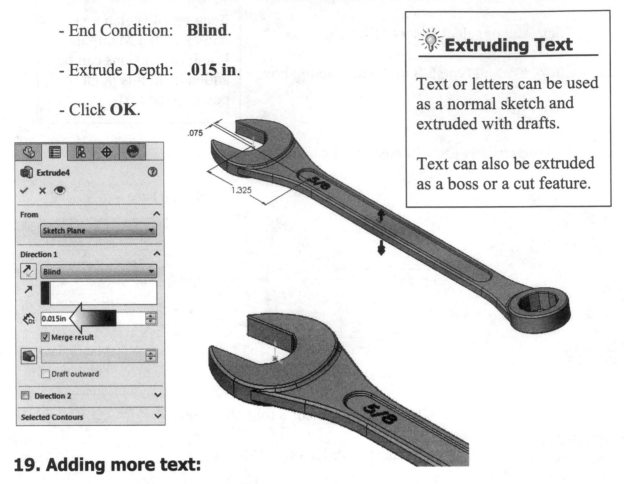 or select **Insert / Boss-Base / Extrude**.

- End Condition: **Blind**.

- Extrude Depth: **.015 in**.

- Click **OK**.

> 💡 **Extruding Text**
>
> Text or letters can be used as a normal sketch and extruded with drafts.
>
> Text can also be extruded as a boss or a cut feature.

19. Adding more text:

- Select the indicated face as sketch plane.

- Click or select **Insert / Sketch**.

- Click and type SPANNER in the Text dialog box.

- Add dimensions to fully position the text.

Sketch Face

More Settings

- Clear Use document's font check box .

- Change Width factor to **150%** .

- Keep Spacing at **100%** .

- Font: **Century Gothic** .

- Style: **Regular**.

- Points size: **14 pt**.

- Click **OK**.

20. Extruding the text:

- Click or select **Insert / Boss-base / Extrude**.

- End Condition: **Blind**.

- Extrude Depth: **.015 in**.

- Click **OK**.

> 🔆 **Text Color**
>
> To change the color of extruded text: Right click the extruded feature (text) from the Feature tree, select Appearances 🔴 and pick a different color.

21. Saving your work:

- Select **File / Save as / Spanner / Save**.

22. Optional:

- To add the same text on the opposite side of the part, repeat from step 17 through step 20.

- Since the mirror option will not work correctly for text, you can either copy the sketch of the text, edit it, re-position, and extrude it again - OR - copy and paste the extruded text and then edit its sketch to reposition.

Questions for Review

Advanced Modeling

1. The Min / Max conditions can be selected from the dimension's properties, under the Leaders tab.
    a. True
    b. False

2. The Mid-Plane extrude type protrudes the sketch profile to both directions equally.
    a. True
    b. False

3. It is sufficient to create a plane at an angle with a surface and an angular dimension.
    a. True
    b. False

4. When sketching a polygon, the number of sides can be changed on the Properties tree.
    a. True
    b. False

5. A 3D solid feature can be mirrored using a centerline as the center of mirror.
    a. True
    b. False

6. Text cannot be used to extrude as a boss or a cut feature.
    a. True
    b. False

7. Extruded text can be mirrored just like any other 3D features.
    a. True
    b. False

8. Text in a sketch can be extruded with drafts, inward or outward.
    a. True
    b. False

7. TRUE
8. TRUE
5. FALSE
6. FALSE
3. FALSE
4. TRUE
1. TRUE
2. TRUE

von Mises (psi)

6.095e+004
5.587e+004
5.079e+004
4.572e+004
4.064e+004
3.556e+004
3.049e+004
2.541e+004
2.033e+004
1.525e+004
1.018e+004
5.101e+003
2.476e+001

<u>Exercise:</u> Circular Text Wraps

1. Opening a part file:

- From the Training Files folder, open an existing part named: **Text Wrap**.

2. Adding Text:

- Select the <u>Top</u> plane and open a new sketch.

- Sketch a circle at Ø**5.500**" and convert it into construction geometry.

- Sketch the other centerlines, trim, then add the dimensions and relations as indicated.

- Click the **Text** command.

- Enter the word: **SolidWorks**.

Symmetric relation (both top and bottom) between the endpoints and the centerlines

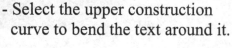

- Select the upper construction curve to bend the text around it.

Select these options

- Click the **Font** button and set the size to **72 points.**

- Select all other options as noted to align the text.

Select this curve

3. Repeating:

- Still working in the same sketch, repeat step 2 and add the text: **Text**.

- Add a Symmetric relation between the endpoints of the construction curve and the vertical centerline.

- Use the same text settings as the last text.

4. Extruding the text:

- Click **Extruded Boss-Base**.

- Change Direction 1 to **Offset From Surface**.

- Enter **.030"** and click **Reverse Offset**.

- Select the **face** as indicated.

- Click **OK**.

Select this face to offset from

5. Saving your work:

- Click **File / Save As**.

- Enter **Circular Text Wrap**.

- Press **Save**.

CHAPTER 4

Sweep with Composite Curves

Sweep 🪱 with Composite Curves

Unlike extruded or revolved shapes, the sweep option offers a more advanced way of creating complex geometry, where a single profile can be swept along 2D guide paths or 3D curves to define the shape.

To create a sweep feature the Sweep Path gets created first, then a single closed sketch Profile.

The Profile will be related to the Sweep Path with a Pierce or a coincident relation.

When the Profile is swept, the Sweep Path and Guide Curves help control the shape and its behaviors such as twisting, tangencies, etc.

The Composite Curve 🗠 option allows multiple sketches or model edges to be jointed into one continuous path to use in sweep features. (The sketches must be connecting with one another in order for the composite curve to work.)

This lesson will guide you through the creation of a helical extension spring where several 2D sketches will be combined with a 3D helix to create one continuous curve.
This curve is called **Composite Curve**.

∅.080

Sweep with Composite Curves
Helical Extension Spring

Dimensioning Standards: **ANSI**

Units: **INCHES** – 3 Decimals

Tools Needed:

Insert Sketch	Line	Circle
Tangent Arc	3 Point Arc	Add Geometric Relations
Dimension	Composite Curve	Sweep

1. Sketching the first profile:

- Select the <u>Front</u> plane from the FeatureManager Tree.

- Click 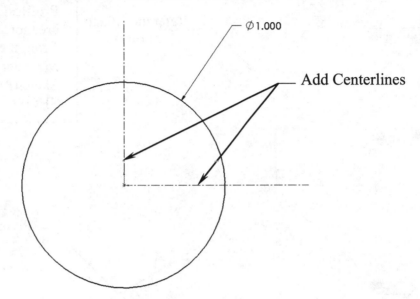 or **Insert / Sketch**.

- Sketch a **Circle** and two centerlines then add the dimension as shown.

Ø1.000

Add Centerlines

2. Converting the circle into a Helix:

- Select **Insert / Curve / Helix / Spiral**.

Defined by:	**Pitch and Revolution**
Pitch:	**.100**
Revolution:	**10**
Starting angle:	**0°**
Taper helix:	**Enabled**
Taper angle:	**10°**

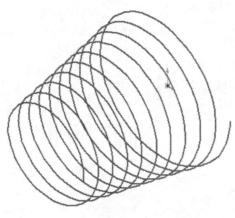

- Click **OK**.

3. Creating a 2-degree plane:

- **Show** the previous sketch (Sketch1).

- Click 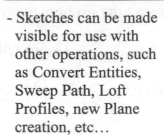 or select **Insert/Reference Geometry/ Plane**.

Reference Plane (Front)

Pivot Line

Show Sketch 👁

- Sketches can be made visible for use with other operations, such as Convert Entities, Sweep Path, Loft Profiles, new Plane creation, etc...
- From the Feature-Manager Tree, click the Sketch1 and select SHOW.

- Select the **Horizontal Centerline** as the Pivot Line.

- Select the **at Angle** option .

- Enter **2.00deg**. for Angle.

- Enable the **Flip** option. (Make sure the new plane leans to the right. Change to the Right view Ctrl+4.)

- Click **OK**.

- Hide the Sketch1.

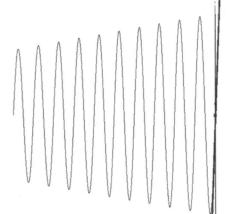

Plane1

2° Planc

4. Sketching the large loop:

- Select the <u>Plane1</u> from the FeatureManager Tree.

- Click 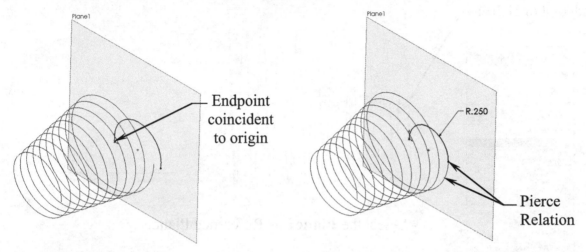 or select **Insert / Sketch**.

- Sketch a **3-point Arc** and add dimension as shown:

- Add a **Pierce** relation between the end point of the Arc and the Helix.

- **<u>Exit</u>** the sketch or select **Insert / Sketch**.

5. Sketching the large hook:

- Select the <u>Right</u> plane from the FeatureManager Tree.

- Click or select **Insert / Sketch**.

- Sketch the profile and add dimension and relations as shown below:

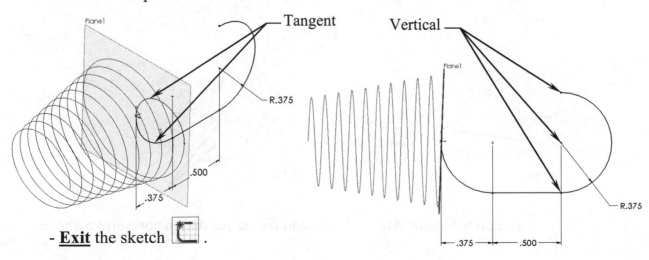

- **<u>Exit</u>** the sketch .

6. Creating a Parallel plane:

- Select the Plane1 from the FeatureManager Tree.

- Click 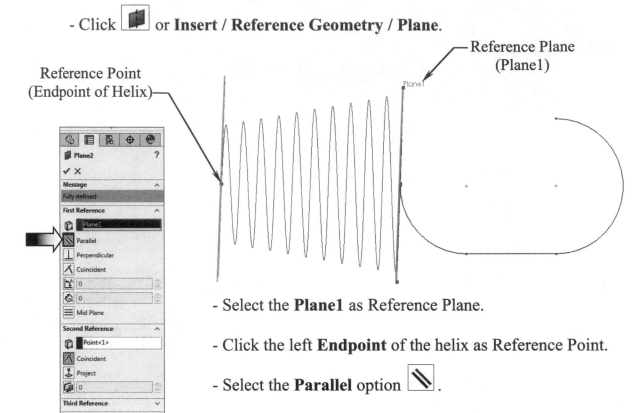 or **Insert / Reference Geometry / Plane**.

- Select the **Plane1** as Reference Plane.

- Click the left **Endpoint** of the helix as Reference Point.

- Select the **Parallel** option .

- Click **OK**.

7. Adding the small loop:

- Select the new plane (Plane2) from the FeatureManager Tree.

- Click or **Insert / Sketch**.

- Sketch a 3-point Arc and add the radius dimension shown above.

- Add a **Pierce** relation between the endpoint of the Arc and the Helix.

- **Exit** the sketch or **Insert / Sketch**.

8. Creating a small hook:

- Select the <u>Right</u> plane from the FeatureManager Tree.

- Click or **Insert / Sketch**.

- Sketch the profile and add the dimensions shown.

.500 — .250

R.250

- Add the relations Vertical and Tangent to the indicated entities.

.250

.500

Vertical

R.250

Tangent

- **Exit** the sketch or click **Insert / Sketch**.

9. Creating a Composite Curve:

- Click 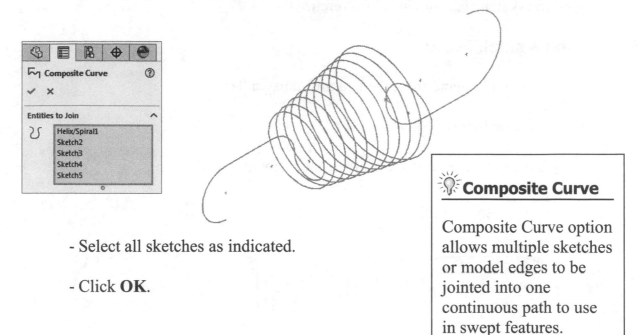 under the Curves button or select: **Insert / Curve / Composite**.

- Select all sketches as indicated.

- Click **OK**.

> 💡 **Composite Curve**
>
> Composite Curve option allows multiple sketches or model edges to be jointed into one continuous path to use in swept features.

NOTE:

*The transitions sketch between the <u>helix</u> and the sketch of the <u>hook</u> are not perfectly tangent. Since we cannot add a fillet between the helix (3D) and the hook (2D) we will try another approach called **Fit Spline**, to smooth out any tangency problem in the model.*

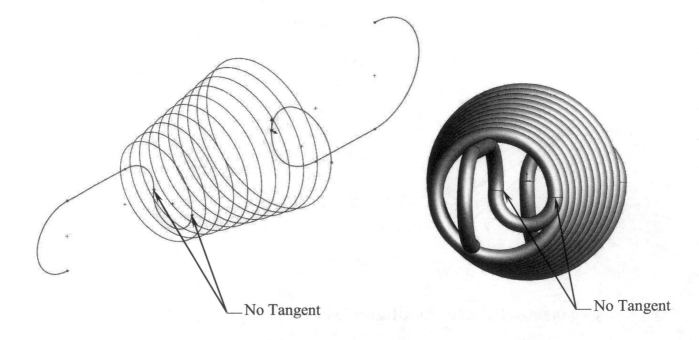

No Tangent

No Tangent

10.   Converting to 3D sketch:

- The Fit Spline command can only be used with sketch entities, not 3D curves. So the next step is to convert the composite curve to a 3D sketch.

- Click **3D Sketch** under the Sketch drop down menu (arrow).

- Select the Composite Curve and click **Convert Entities** .

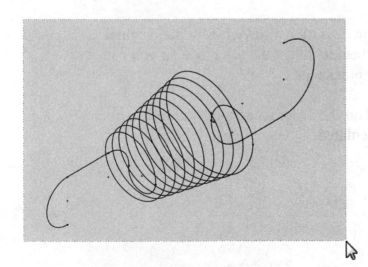

Curve to 3D sketch

- The composite curve is converted to a 3D sketch.

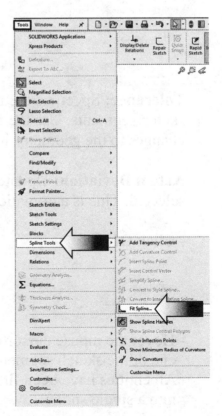

- Box-select the entire 3D sketch and click:

Tools / Spline Tools / Fit Spline .

- Select / enter the following:

        * Constraint.
        * Tolerance: .003in.
        * Inflection Points.
        * Minimum Radius.
        * Curvature Comb.

> **☼ Fit Spline tool** ⌐
>
> The Fit Spline tool is used to fit sketch segments to a spline. Fit splines are parametrically linked to underlying geometry so that changes to the geometry update the spline.

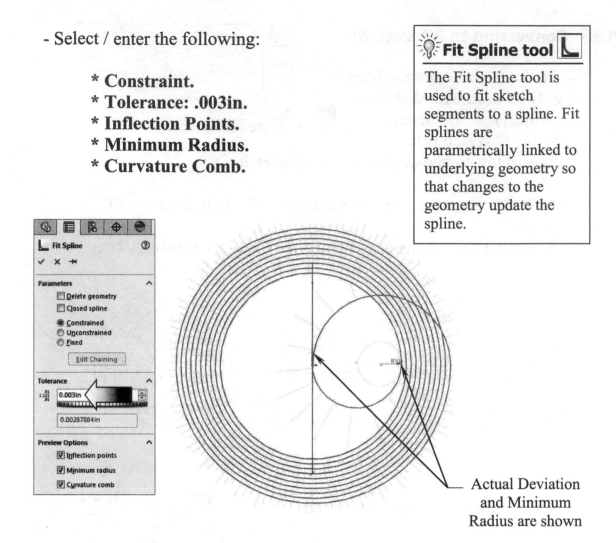

Actual Deviation and Minimum Radius are shown

Tolerance: Specifies the maximum deviation allowed from the original sketch segments. Use the thumbwheel to adjust the tolerance so you can see changes to the geometry in the graphics area.

Actual Deviation: Updates based on the Tolerance value and the geometry selected. This is automatically calculated.

- Click **OK**.

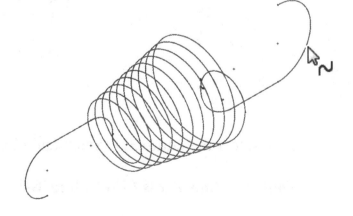

- Hover the mouse cursor over one of the sketch segments. All entities have been fitted into a <u>single spline</u>.

11. **Sweeping the profile along the path:**

- Click on the Features toolbar or select: **Insert / Boss-Base / Sweep**.

- Select the **Circular Profile** option and enter **.080in** for profile diameter ⊘.

- Select the **Composite Curve** as the sweep path.

- Click **OK**.

Before Fit Spline

After Fit Spline

12. **Saving your work:**

- Click **File / Save As /**

- Enter **Helical Extension Spring** for the file name and click **Save**.

Other Spring Examples:

Conical Spring

V-Shaped - 1 Rev. Closed

Ground-End

Closed-End

Circular Expanded

V-Shape - 3 Rev. Open

Questions for Review

Sweep with Composite Curve

1. Beside the Pitch and Revolution option, a helix can be defined with Pitch and Height.
   a. True
   b. False

2. It is sufficient to create an Offset Distance plane using a reference plane and a distance.
   a. True
   b. False

3. The sweep profile should have a Pierce relation with the sweep path.
   a. True
   b. False

4. Several sketches or model edges can be combined to make a Composite curve.
   a. True
   b. False

5. A Composite curve cannot be used as a sweep path.
   a. True
   b. False

6. The composite curve combines all sketches and model edges into one continuous curve, even if they are not connected.
   a. True
   b. False

7. In a sweep feature, SOLIDWORKS allows only one sweep path, but multiple guide curves can be used.
   a. True
   b. False

8. Several sketch profiles can be used to sweep along a path.
   a. True
   b. False

1. TRUE	2. TRUE
3. TRUE	4. TRUE
5. FALSE	6. FALSE
7. TRUE	8. FALSE

Exercise: Circular Spring – 180deg.

Front Plane

R3.000

Origin

Right Plane

.250

1. Sketching the Sweep Path:

- Select the <u>Front</u> plane and open a new sketch.

- Sketch an **Arc** as shown and add a Horizontal relation between the left and the right endpoints, then add a radius dimension.

- **Exit** the sketch.

2. Sketching the Sweep Profile:

- Select the <u>Right</u> plane and open a new sketch.

- Sketch a horizontal **Line** towards the right.

- Add a **.250 in**. dimension.

- **Exit** the sketch.

3. Creating a Swept <u>Surface</u>:

- Click or select **Insert / Surface / Sweep**.

- Select the horizontal **Line** to use as the Sweep Profile.

- Select the **Arc** as the Sweep Path.

More...

- Expand the **Options** dialog box.

- Select **Twist Along Path**, under Orientation / Twist Type.

- For Define By: Select **Turns**.

- For number of Turns: Enter **30**.

- Click **OK**.

4. Sketching the Wire-Diameter:

- Select the <u>Right</u> plane and open a new sketch.

- Sketch a **Circle** at the right end of the swept surface.

- Add a **Ø.125 in**. diameter dimension.

- **Exit** the sketch.

5. Creating a Swept Boss-Base: (Solid)

- Click or select **Insert / Bose-Base / Sweep**.

- Select the **Circle** to use as the Sweep Profile.

- For Sweep Path, select the **Edge** of the Swept-Surface.

- Click **OK**.

6. Hide the Swept-Surface:

- Right click over the Swept-Surface and select **Hide**.

7. Save your work:

- Select **File / Save As.**

- For file name, enter **Expanded Circular Spring**.

- Click **Save**.

Right click
& HIDE

Using Variable Pitch

Sweep with Variable Pitch Helix

- In a Sweep feature, there is only one Sweep Profile, one Sweep Path, and one or more Guide Curves.

- The Sweep Profile describes the feature's cross-section; the Sweep Path helps control the twisting and how the Sweep Profile moves along the path.

- The Sweep path can either be a 2D or a 3D sketch, the edges of the part, or a Composite Curve.

- Beside the Pitch and Revolution option, the Helix / Spiral command offers other options to create more advanced curves such as:

Region parameters:				
	P	Rev	H	Dia
1	0.115in	0	0in	1in
2	0.115in	1.5	0.1725	1in
3	0.375in	6.5	1.3975	1in
4	0.25in	11.5	2.96in	1in
5	0.115in	12.5	3.1425	1in
6	0.115in	14	3.315i	1in
7				

        * Height and Revolutions.
        * Height and Pitch.
        * Spiral.
        * Constant Pitch.
        * Variable Pitch.

- We will take a look at the Variable Pitch option in this lesson and learn how a helix with variable pitch is created using a table, to help control the changes of the dimensions.

- To create the flat ground ends, an extruded cut feature is added at the end of the process.

Using Variable Pitch
Multi-Pitch Spring with Closed Ends

Dimensioning Standards: **ANSI**
Units: **INCHES** – 3 Decimals

Tools Needed:

Insert Sketch	Helix / Spiral	Add Geometric Relations
Dimension	Base/Boss Sweep	Extruded Cut

1. Creating the base sketch:

- Select the <u>Front</u> plane from the FeatureManager tree.

- Click 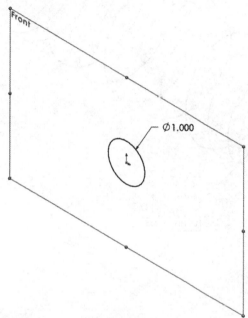 or select: **Insert / Sketch**.

- Sketch a **Circle** centered on the origin.

- Add a diameter dimension of **1.000"**.

Ø1.000

2. Creating a helix with Variable Pitch:

- Click or select **Insert / Curve / Helix-Spiral**.

- Under Define By, select:
  Pitch and Revolution.

- Under Parameter, select:
  Variable Pitch.

- Under Region Parameters,
  enter the values for the **Pitch**,
  Revolutions, and **Diameters**.
  (Ignore the Height column, SOLIDWORKS
  will fill in the values automatically.)

Region parameters:

	P	Rev	H	Dia
1	0.115in	0	0in	1in
2	0.115in	1.5	0.1725	1in
3	0.375in	6.5	1.3975	1in
4	0.25in	11.5	2.96in	1in
5	0.115in	12.5	3.1425	1in
6	0.115in	14	3.315i	1in
7				

- Set Angle to **0deg** and **Clockwise** direction.

- Your Variable Pitch helix should look like the image shown below.

- Click **OK**.

- Press **Control + 4** to view the Variable-Pitch Helix from the right side.

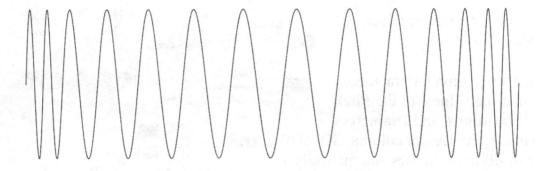

3. Sweeping the profile along the path:

- Click or select: **Insert / Boss-Base / Sweep**.

- Select the **Circle** Profile option.

- Enter **.100in** for profile diameter.

- For sweep path, select the **Helix** either from the feature tree or from the graphics area.

- Click **OK**.

- The resulting Swept feature.

4. Creating a trimmed sketch:

- Select the <u>Top</u> plane from the Feature-Manager tree and open a new sketch.

- Sketch a **Rectangle** and add a **Midpoint** relation between the line on top and the origin.

- Add a width and a height dimension. The sketch should now be fully defined.

5. Extruding a cut:

- Click or select: **Insert / Cut / Extrude**.

- Set the Direction 1 to **Through All Both**.

- Enable the **Flip Side to Cut** checkbox.

- Click **OK**.

6. Saving your work:

- Click **File / Save As / Variable Pitch Spring / Save**.

Questions for Review

Sweep with Composite Curve

1. Multiple Sweep Profiles can be used in a sweep feature.
   a. True
   b. False

2. Multiple Sweep Paths can be used in a sweep feature.
   a. True
   b. False

3. Only one Sweep Profile and one Sweep Path can be used in a sweep.
   a. True
   b. False

4. A Helix can be defined by:
   a. Pitch and Revolution
   b. Height and Revolution
   c. Height and Pitch
   d. Spiral
   e. All of the above

5. Several connected Helixes can be combined into one single Composite Curve.
   a. True
   b. False

6. The Sweep Profile sketch should be related to the Sweep Path using the relation:
   a. Perpendicular
   b. Parallel
   c. Coincident
   d. Pierce

7. The Sweep Path controls the twisting and how the Sweep Profile moves along.
   a. True
   b. False

8. The Edges of the part can also be used as the Sweep Path.
   a. True
   b. False

7. TRUE
8. TRUE
5. TRUE
6. D
3. TRUE
4. E
1. FLASE
2. FALSE

<u>Exercise:</u> Projected Curve & Composite Curve

1. Create the part based on the drawing as shown.
2. Dimensions are in inches, 3 decimal places.
3. Focus on Projected Curve & Composite Curve options.

Ø.172 THRU

Ø.980

A

A

(15°)

B

(32°)

SECTION A-A

.058

.212

1.75 REVOLUTION
.073 PTICH

.595

.356

.015

.073

.029

R.029

DETAIL B

4. Follow the instructions on the following pages, if needed.

1. Opening a part document:

- From the Training Files folder, locate and open the part document named:
Project and Composite Curves.

- This is an actual die to form the shape of a catheter. We will create the groove that wraps around this die, using the Project and Composite Curve options.

2. Creating a helix:

- Select the <u>Top</u> plane and open a new sketch.

- Select the **bottom edge** (or top edge) and press **Convert- Entity**.

- Click 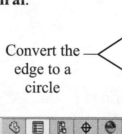 or select:
Insert / Curve / Helix-Spiral.

Convert the edge to a circle

- Enter the following:

   * **Constant Pitch**.

   * Pitch: **.073"**

   * **Reverse Direction**.

   * Revolutions: **1.75**

   * Start Angle: **15deg**.

   * **Counterclockwise**.

- Click **OK**.

- Notice the helix starts at a 15° angle? This way the end of the helix will match up with the bend radius in the next step.

3. Sketching the upper transition:

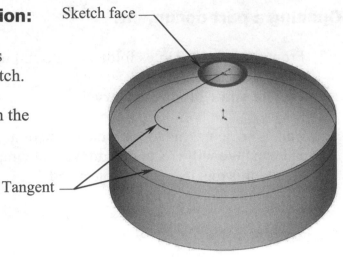

Sketch face

- Select the small <u>upper face</u> as indicated and open a new sketch.

- Sketch a **Line** that starts from the origin and connects with a **Tangent Arc**.

Tangent

- Add a **Tangent** relation between the arc and the circular edge of the part.

- Sketch a **centerline** that starts from the origin and connects to the right endpoint of the arc.

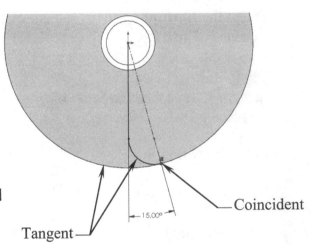

- Add a **Coincident** relation between the right endpoint of the arc and the circular edge of the part, then add a **15°** angular dimension.

Coincident

- The sketch should be fully defined at this point. <u>Exit</u> the sketch.

15.00°

Tangent

4. Creating a projected Curve:

- Click [icon] or select: **Insert / Curve / Projected**.

Select 2 faces

- Select the **Sketch on Faces** option.

- For Sketch to Project, select the **Sketch3**.

- For Projection Faces, select the **2 faces** as noted. Click Reverse.

- Click **OK**.

5. Adding a sketch line:

Sketch face

- We want the cut to start from the origin; a sketch line is needed to connect the projected curve to the origin.

- Sketch a line that starts from the origin to the endpoint of the projected curve.

- **Exit** the sketch.

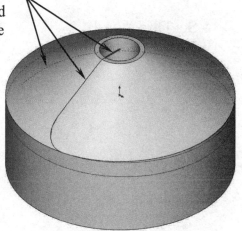

Add a line

6. Creating a Composite Curve:

- Click or select: **Insert / Curve / Composite**.

- Select either from the Feature tree or from the graphics: the **Helix**, the **Curve1** and the **Sketch4** (the line).

- Click **OK**.

Select the Helix the Curve1 and the Sketch line

7. Creating a swept cut:

- Click 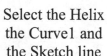 or select: **Insert / Cut / Sweep**.

- Select the **Circular Profile** option.

- For Profile Diameter enter **.058in**.

- For Sweep Path, select the **Composite Curve**.

- Click **OK**.

8. Removing the undercut:

- If the swept feature stopped short, we'll need to clean it up. Select the face as noted and open a new sketch.

Sketch face

Convert entities

- Convert the face into a sketch.

- Click or select: **Insert / Cut / Extrude**.

- Set Direction 1 to: **Through All**.

- Click **OK**.

9. Saving your work:

- Click **File / Save As**.

- Enter **Project and Composite Curves** for the name of the file.

- Press **Save**.

CHAPTER 5

Advanced Modeling with Sweep & Loft

Advanced Modeling with Sweep & Loft

- The Sweep option creates a solid, thin, or surface feature by moving a single sketch profile along a path and guiding with one or more guide curves.

- In order to create a sweep feature properly, a set of rules should be taken into consideration:

- o The Sweep option uses only one sketch profile and it must be a closed non-intersecting contour for a **solid** feature.
- o The sketch profile can be either closed or open for a **surface** feature.
- o Only one path is used in a sweep and it can be open or closed.
- o One or more guide-curves can be used to guide the sketch profile.
- o The sketch profile must be drawn on a new plane starting at the end point of the path.

- The Loft option creates a solid, thin, or surface feature by making a transition between the sketch profiles.

- Keep in mind the following requirements when creating a loft feature:

- o The Loft option uses multiple sketch profiles that must be closed and non-intersecting for a **solid** feature.
- o The sketch profiles can be either closed or open for a **surface** feature.
- o Use the Centerline Parameter option to guide the profiles from the inside.
- o Use Guide Curves option to guide the sketch profiles from the outside.
- o The Guide Curves can be either a 2D or a 3D sketch.

Advanced Modeling with Sweep & Loft
Water Pump Housing

View Orientation Hot Keys:

Ctrl + 1 = Front View
Ctrl + 2 = Back View
Ctrl + 3 = Left View
Ctrl + 4 = Right View
Ctrl + 5 = Top View
Ctrl + 6 = Bottom View
Ctrl + 7 = Isometric View
Ctrl + 8 = Normal To
         Selection

Dimensioning Standards: **ANSI**

Units: **INCHES** – 3 Decimals

Tools Needed:

 Insert Sketch   Split Entities   Add Geometric Relations

 Rib   Plane   Revolved Boss/Base

 Extruded Boss/Base   Swept Boss/Base   Lofted Boss/Base

Understanding the Draft Options

- Drafts are normally required in most plastic injection molded parts to ensure proper part removal from the mold halves.
- The Draft option in SOLIDWORKS adds tapers to the faces using the angles specified by the user.
- Drafts can be inserted in an existing part or added to a feature while being extruded.
- Drafts can be applied to solid parts as well as the surface models.
- There are several types of draft available:

* Neutral Plane
* Parting Line
* Step Draft

Neutral Plane Draft

Split Line Draft

Step Draft

1. Opening a part document:

- From the Training Files folder, locate and <u>open</u> the part document named: **Water Pump Sketch**.

- The sketch was created ahead of time to help focus on the key features of this lesson: Sweep and Loft.

- <u>**Edit**</u> the Sketch1.

2. Extruding the Base with Draft:

- Click **Extruded Boss/Base.**

- End Condition: **Blind**.

- Depth: **1.00 in**.

- Draft: **7 deg. inward**.

- Click **OK**.

3. Sketching the upper Inlet Port:

- Select the <u>Front</u> plane and open a new sketch.

- Sketch the profile as shown.

- Add the dimensions and relations as indicated.

- Add a **Vertical Centerline** from the Origin and use it as the center of the revolve in the next step.

Virtual Diameter: Measured from the centerline to the endpoint of the line.

Parallel

Collinear

4. Revolving the upper Inlet Port:

- Click **Revolve Boss/Base** or Select: **Insert / Features / Boss-Base / Revolve** from the drop-down menus.

- The vertical centerline should be selected automatically to use as the Axis of Revolution.

- Revolve Direction: **Blind**.

- Revolve Angle: **360 deg**.

- Click **OK**.

5. Adding the .500" Fillets:

- Click **Fillet** or select **Insert / Features / Fillet-Round.**

- Enter **.500 in.** for radius.

- Select the **7 edges** as indicated.

- Click **OK**.

Select
7 edges

6. Adding the .275" Fillets:

- Click **Fillet** or select **Insert / Features / Fillet-Round.**

- Enter **.275 in.** for radius value.

- Select the **upper edge** of the base.

- Make sure the option **Tangent Propagation** is **Enabled** so that the fillet can propagate itself to all connecting edges.

- Click **OK**.

Select the
Upper edge

7. Creating the 1st Offset-Distance Plane:

- Click **Plane** or select **Insert / Reference Geometry / Plane.**

- From the FeatureManager tree, select the **Front** plane to offset from.

- Enter **3.000 in**. for distance.

- Place the new plane on the **right side** of the front plane.

- Click **OK.**

8. Creating the 2nd Offset-Distance Plane:

- Click **Plane** or select **Insert / Reference Geometry / Plane.**

- Select the <u>Front</u> plane again from the FeatureManager tree to offset from.

- Enter **5.000 in**. for offset distance.

- Place the new plane also on the **right side** of the front plane.

- Click **OK.**

9. Creating the 3rd Offset-Distance Plane:

- Click **Plane** or select **Insert / Reference Geometry / Plane.**

- Select the <u>Front</u> plane once again from the FeatureManager tree to offset from.

- Enter **6.000 in**. for offset distance.

- Place the new plane also on the **right side**.

- Click **OK**.

10. Sketching the 1st loft profile:

- Select the <u>Front</u> plane and open a new sketch .

- Sketch a **Rectangle** that is just **.125 in**. above the bottom edge of the part.

- Add the dimensions and relations as shown to fully define the sketch.

Symmetric relation

1.750

2.500

.125

- <u>Exit</u> the sketch.

11. Sketching the 2nd loft profile:

- Select the <u>Plane1</u> and open a new sketch .

- Sketch another **Rectangle** as shown.

- Add the dimensions and relations needed to fully define the sketch.

- **Exit** the sketch.

Symmetric relation

Plane3

1.250

.375

2.000

12. Sketching the 3rd loft profile:

- Select the <u>Plane2</u> and open a new sketch .

- Sketch a **Circle** just above the Origin as shown below.

- Use the **Split-Entities** command and split the circle into **4 segments**.

- Add Vertical and Horizontal relations between the split points.

- The Split Entities command is used to split the entities in each sketch to an even number of connecting points to help control the twisting in a loft feature.

- Add the dimensions and relations needed to fully define the sketch.

- **Exit** the sketch.

Split Entities

Ø1.250

Coincident

Plane3

1.000

13. Sketching the 4ᵗʰ loft profile:

- Select the <u>Plane3</u> and open a new sketch .

- Convert the circle from the previous sketch;
  this creates an On-Edge relation between
  the 2 circles and they will update at
  the same time when the first
  circle is changed.

Convert from the
previous sketch…

Plane3

- **Exit** the sketch.

14. Creating a loft feature:

- Click Loft or select: **Insert /
  Boss-Base / Loft** from the drop
  down menus.

- Select the **4 sketch profiles** in
  the graphics area.

  *(Since there are no guide curves
  to help control the loft, the
  profiles should be selected
  from the same side each
  time to prevent them
  from twisting.)*

- For clarity, right click
  in the yellow shaded
  area and select the
  following options:

  * Opaque Preview
  * Clear Meshed Faces

- Click **OK**.

Select the 4 sketch
profiles, use the same
connecting point of
each profile…

Profile(Sketch6)

15. Creating the mounting bosses:

- Select the <u>bottom face</u> and open a new sketch.

- Sketch **5 Circles** as shown*.

* <u>Avoid</u> the hidden entities.

(When sketching, your circles may snap to some of the hidden edges of the model causing an over defined error when adding the dimensions.
To overcome this, hold the control key every time a hidden edge highlights; it will cancel the Auto-Relation snapping).

Vertical
Vertical

.275

Ø.750

Sketch Face

Tangent with Edge of part (5 places.)

2.100   3.450

- Add a **Tangent** relation for each circle, to the outer edge of the part.

- Add an **Equal** relation to all 5 circles.

- Add dimensions to fully position the 5 circles.

16. Extruding the 5 mounting bosses:

- Click **Extrude Boss-Base** or select: **Insert / Extrude / Boss-Base** from the drop-down menus.

- Set the following:

- End Condition: **Blind**

- Depth: **1.250 in**.

- Draft: **1 deg. Inward.**

- Click **OK**.

Extrude2

From
Sketch Plane

Direction 1
Blind

1.250in
☑ Merge result

1.00deg
☐ Draft outward

☐ Direction 2

Selected Contours

.275   Ø.750   2.100   3.450

17. Sketching the rear Inlet Port:

- Select the <u>upper face</u> and open a new sketch.

- Sketch the profile of the Inlet at a **30° angle**; one end of the profile is locked onto the Origin.

- Add dimensions and other relations to fully position the sketch.

- There are more than one centerline in this sketch. Select the 30° centerline before clicking the revolve command.

Perpendicular

Sketch Face

.500

4.000   30.00°   .750

.375

Revolve centerline

18. Revolving the Rear Inlet Port:

- Click **Revolve** or select:
  Insert / Boss-Base / Revolve.

- Select the 30° centerline and set the following:

- Revolve Type: **Blind**.

- Revolve Angle: **360 deg**.

- Merge Result: **Enabled**

- Click **OK**.

19. Adding the 1st Face Fillet:

- Click **Fillet** or select: **Insert / Features / Fillet-Round**.

- Select the **Face Fillet** button.

- Enter **.250 in.** for radius.

- For **Face Set 1**, select the **upper face** of the lofted feature.

- For **Face Set 2**, select the **side face** of the lofted feature.

- Click **OK**.

> 💡 **Face Fillet**
>
> A Face Fillet is used to create a blend between non-adjacent, non-continuous faces.

Face Set 1 — Face Set 1: 0.25in

Face Set 2

—— Face Set 2 ——

20. Adding the 2nd Face Fillet:

- Click **Fillet** or select: **Insert / Features / Fillet-Round.**

- Select the **Face Fillet** button again.

- Enter **.250 in.** for radius value.

- For **Face Set 1**, select the **upper face** of the lofted feature.

- For **Face Set 2**, select the **side face** of the lofted feature.

- Click **OK**.

Face Set 1 —

Face Set 1: 0.25in

Face Set 2

—— Face Set 2

21. Adding the 3rd Face Fillet:

- Click **Fillet** or select:
Insert / Features / Fillet-Round.

- Click **Face Fillet**.

- Enter **.150 in.** for radius value.

- For **Face Set 1**, select the **lower face** of the lofted feature.

- For **Face Set 2**, select the **side face** of the lofted feature.

Face Set 1

Face Set 2

- Click **OK**.

22. Adding the 4th Face Fillet:

- Click **Fillet** or select:
Insert / Features / Fillet-Round.

- Click **Face Fillet**.

- Enter **.150 in.** for radius value.

- For **Face Set 1**, select the **lower face** of the lofted feature.

- For **Face Set 2**, select the **side face** of the lofted feature.

- Click **OK**.

Face Set 1

Face Set 2

23. Mirroring the rear Inlet Port:

- Click **Mirror** or select: **Insert / Pattern Mirror / Mirror.**

- For Mirror Face/Plane, select the **Right** plane from the FeatureManager tree.

- For Features to Mirror, select the **Rear Inlet Port** either from the graphics area or from the feature tree.

- Click **OK**.

24. Adding the .175" Fillets:

- Click **Fillet** or select: **Insert / Features / Fillet-Round.**

- Select the **Constant Size** fillet option.

- Enter **.175 in.** for radius value.

- Select the edges as noted to add the fillets.

- The option Tangent-Propagation should be enabled by default.

- Click **OK** ✅.

Select edges to fillet

Radius: 0.175in

25. Shelling the part:

- Click **Shell** or
  select: **Insert /
  Features / Shell.**

- Under the Parameters
  section, enter **.080 in**. for
  wall thickness.

- Select a total of **10
  faces** to remove.

- Click **OK**.

10 faces to remove

26. Adding the .0625" Fillets:

- Click **Fillet** or select:
  **Insert / Features /
  Fillet-Round.**

- Enter **.0625 in**.
  for radius value.

- Select the **edges**
  as noted to add
  the fillets.

- Click **OK**.

Edges to fillet
(outer edges)

27. Adding a Rib:

- Select the <u>Front</u> plane from the FeatureManager tree

  and open a new sketch .

- Sketch a **Line** as shown.

- Add the coincident relations between the endpoints of the line and the edges of the part.

- Add the location dimensions shown.

> 💡 **Rib Features**
>
> A rib is an extruded feature which adds material of a specified thickness in a specified direction. Drafts can also be added to the faces of the rib.

Coincident with edges

3.500

2.500

28. Extruding the Rib:

- Click **Rib** or select: **Insert / Features / Rib.**

- Select **Both Directions** under the Thickness section.

- Enter **.275 in**. for the thickness of the rib.

- Enable the **Draft** option and enter **1.00 deg**.

- Enable the **Draft Outward** check box.

- Click **OK**.

29. Creating a Full-Round fillet:

- A Full-Round creates fillets that are tangent to three adjacent face sets.

- Click **Fillet** or select: **Insert / Features / Fillet-Round.**

- Select the **Full Round** fillet option.

- For Face 1, select the **left face** of the Rib.

- For Face 2, select the **middle face** of the Rib.

- For Face 3, select the **right face** of the Rib.

- Click **OK**.

Face 1 – Left Side of Rib

Face 2 – Middle of Rib

Face 3 – Right Side of Rib

30. Mirroring the Rib:

- Click **Mirror** or select: **Insert / Pattern Mirror / Mirror.**

- For Mirror Face/Plane, select the **Right** plane.

- For Features to Mirror, select the **Rib** and its **fillet**.

- Click **OK**.

31. Adding the .025" fillets:

- Click **Fillet** or select: **Insert / Features / Fillet- Round.**

- Select the **Constant Radius** fillet option.

- Enter **.025 in**. for radius value.

- Select the edges as indicated to add the fillets.

- Enable the Tangent-Propagation checkbox.

- Click **OK**.

Select edges
(Inner Edges)

32. Adding the .175" fillets:

- Click **Fillet** or select: **Insert / Features / Fillet-Round.**

- Click **Constant Radius**.

- Enter **.175 in**. for radius value.

- Select the **edges** of the 2 ribs as indicated.

- Click **OK**.

Select the edges of
the Rib to add fillets

Radius: 0.175in

33. Saving your work:

- Click **File / Save As**.

- Enter **Water Pump Cover** as the name of the file.

- Click **Save.**

CHAPTER 6

Loft vs. Sweep

Loft vs. Sweep

- The Loft and the Sweep commands are normally used to create advanced, complex shapes. The differences between the two are:

   • Sweep uses a single sketched profile to sweep along a path and is controlled by one or more guide curves.

Sweep uses one profile, one path and multiple guide curves

   • Loft uses multiple sketched profiles to loft between the sections and is controlled by 1 or more guide curves or 1 Centerline Parameter.

Loft uses multiple profiles, one centerline parameter, and/or multiple guide curves

- In order to create a solid feature, each sketch profile must be a single, closed, and non-intersecting shape.

- The guide curves can be either a 2D sketch or a 3D curve.

- The sweep path and guide curves must be related to the sketch profiles with either a Coincident or Pierce relation.

- The loft profiles should have the same number of entities or segments. The Split-Entities commands can be used to split the sketch entities and add the necessary connectors to help control the loft more accurately.

Loft vs. Sweep
Water Meter Housing

Dimensioning Standards: **ANSI**

Units: **INCHES** – 3 Decimals

Tools Needed:

Insert Sketch	Line	Split Entities
Sketch Fillet	Mirror	Add Geometric Relations
Dimension	Base/Boss Extrude	Extruded Cut
Plane	Base/Boss Sweep	Base/Boss Loft

1. Sketching the 1st Loft Profile:

- Select the <u>Top</u> plane from the FeatureManager tree.

- Click or select **Insert / Sketch**.

- Sketch a **rectangle** centered on the Origin and add the dimensions shown.

Mid-Point relation

6.000

5.000

R.500

- **Exit** the sketch.

2. Creating a new plane:

- Click or select **Insert / Reference Geometry / Plane**.

- Select the <u>Top</u> reference plane from the FeatureManager tree.

- Select **Offset Distance** option and enter **3.000** in.

- Click **OK**.

New Plane

3. Sketching the 2nd Loft Profile:

- Select the <u>new plane</u> (**Plane1**) and open a new sketch .

- Sketch a Circle and split it into 8 segments (**Tools / Sketch Tools / Split**).

- Add dimensions and the vertical / horizontal relations between the split points.

Split the circle into 8 segments, and add the vertical & horizontal relations to the points.

85.00°

Ø7.000

120.00°

Vertical

Horizontal

(Click the 3 points as noted to create the angle dim).

- **Exit** the Sketch .

4. Creating the Lofted Base feature:

- Click or select **Insert / Boss-Base / Loft**.

- Profiles: Select the **2 sketches** as indicated.

Box Selection
Lasso Selection
SelectionManager
Select Other

Clear Selections

Zoom/Pan/Rotate ▶

✓ OK
✗ Cancel

Merge Tangent Faces
Close Loft

Show All Connectors
Hide All Connectors

Mesh Preview ▶
Show All Curvature Combs
Zebra Stripes Preview
Transparent Preview

Customize Menu

Loft1

Profiles
Sketch2
Sketch1

Start/End Constraints
Guide Curves
Centerline Parameters
Sketch Tools
Options
☑ Merge tangent faces
☐ Close loft
☑ Show preview
☑ Merge result
☐ Curvature Display

- Click **OK**.

Right click and pick:
Show All Connectors

Profile(Sketch1)

5. Constructing the Inlet's 1ˢᵗ Loft Profile:

- Select the <u>Front</u> plane and open a new sketch.

- Sketch a **Rectangle**, add dimensions and sketch fillets.

6. Creating an Offset Distance plane at 5.500":

- Click or select **Insert / Reference Geometry / Plane**.

- Select the **Front** reference plane from the FeatureManager tree.

- Select **Offset Distance** option and enter **5.500 in**. for distance.

- Click **OK**.

New Plane

- **Exit** the Sketch.

7. Constructing the Inlet's 2nd Loft Profile:

- Select the <u>Plane2</u> and open a new sketch .

- Sketch a **Circle** and add Dimensions to fully define.

- Click **Split Entities** and split the circle into 8 segments.

Split the circle into 8 segments and add the vertical & horizontal relations between the points.

Ø2.000

Plane2

Vertical

1.500

90.00°

135.00°

Ø2.000

Plane2

90.00°

135.00°

1.500

- **Exit** the Sketch .

8. Creating the Inlet feature:

- Click or select **Insert / Boss-Base / Loft**.

- For profiles: Select the **2 sketches** from the graphics area.

- Show all connector points to ensure a smooth transition between the 2 profiles. Enable the **Merge Tangent Faces** option.

- Click **OK**.

Right click & Show-All-Connectors

Profile(Sketch4)

9. Creating an Offset Distance plane from the Right:

- Click or select **Insert / Reference Geometry / Plane.**

- Select the **Right** reference plane from the FeatureManager tree.

- Select **Offset Distance** and enter **4.000 in**. for distance; place it on the right.

- Click **OK**.

New Plane

10. Constructing the Outlet's 1st Loft profile:

- Select the new plane (Plane3) and open a new sketch.

- Sketch a **Circle** split it into 8 segments. Add dimensions as noted.

Split the circle into 8 segments and add the vertical & horizontal relations to the split points.

75.00°   Ø2.000

1.500

120.00°

6.000

- Add the Vertical and Horizontal relations to the endpoints of the arcs to fully define them.

- **Exit** the Sketch .

11. "Re-Using" the previous sketch*:

- Expand the feature **Loft2** from the FeatureManager tree (click the **+** sign).

- Locate **sketch3** (the sketch of the Rectangle), click it and select **SHOW** .

Show Sketch*

* The previous sketch is made visible in the graphics area to be used again in the next loft operation.

12. Creating a plane Parallel:

- Click or select **Insert / Reference Geometry / Plane**.

- For 1st reference, select the **Top** reference plane from the FeatureManager tree.

- Click the **Parallel** option (arrow).

- For 2nd reference, click the **Center Point** of the circle as noted.

- Click **OK**.

Reference Point

Reference Plane
(TOP)

- The new plane is
  created (Plane4).

13. Constructing the Centerline Parameter:

- Select the <u>new plane</u> (Plane4) and open a new sketch ▣ .

- Switch to the Top view orientation ▣ (Ctrl+5).

- Sketch **2 Lines** and add a **Sketch Fillet** as shown below.

- The end points of the lines must be coincident of Pierced to the centers of the other sketches.

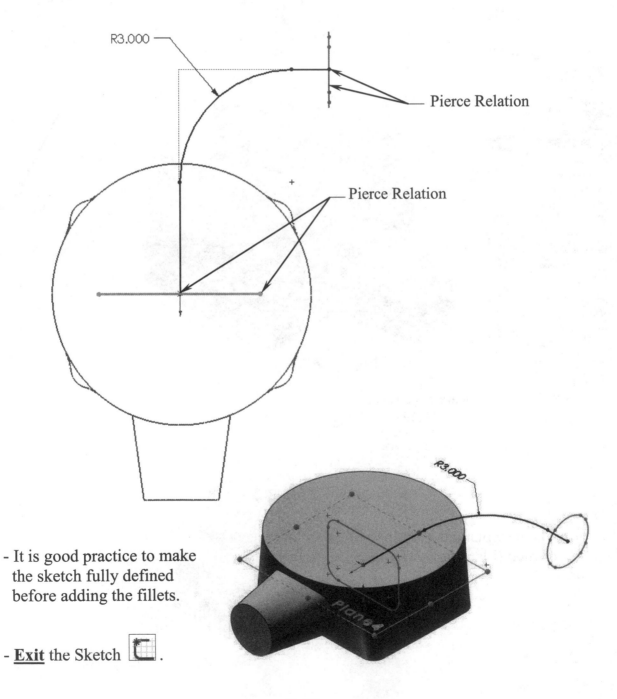

R3.000

Pierce Relation

Pierce Relation

- It is good practice to make the sketch fully defined before adding the fillets.

- **Exit** the Sketch ▣ .

14. Creating the Outlet loft feature:

- Click or select **Insert / Boss / Loft**.

- Select the **2 Sketch Profiles** (Rectangle and Circle) from the graphics area.

- Expand the **Centerline Parameters** option and select **Sketch6** from either the graphics area or from the FeatureManager tree.

- When the Centerline Parameter is used to guide the loft, the sketch planes of all the intermediate sections will be rotated normal to the centerline.

- After the preview appears, check the connectors to ensure a proper loft transition.

- Click **OK**.

> 💡 **Centerline Parameters**
>
> If the number of entities in each sketch is the same, a "Centerline Parameter Sketch" can be used instead of the guide curves.

Centerline Parameter

Loft Profiles

Right click on one of the connectors and select: Show All Connectors

15. Adding .250" fillet to the Bottom:

- Click or select **Insert / Features / Fillet-Round**.

- Enter **.250 in**. for radius value and select the **bottom edges** as indicated.

- Click **OK**.

Radius: 0.25in

Edges to fillet

16. Add .175" fillets to the sides:

- Click or select **Insert / Features / Fillet-Round**.

- Enter **.175 in**. for Radius and select <u>all</u> side edges as noted.

- Click **OK**.

Select <u>all</u> edges

Radius: 0.175in

17. Shelling the part:

- Click or select **Insert / Features / Shell**.

- Enter **.175 in.** for wall thickness and select the **3 faces** as noted.

- Click **OK**.

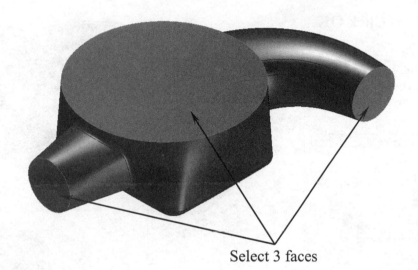

Select 3 faces

18. Creating the 1st Mounting bracket:

- Select the <u>Face</u> as noted and open a new sketch ▢ .

- Sketch the profile as shown below; use Convert Entities ▢ where applicable.

- Add Dimensions ▧ and Relations ⊥ needed to fully define the sketch.

R1.375

Convert
Circle

R.750

∅.500

5.000

The center of R1.375 is coincident with the center of the circle.

Sketch Face

19. Extruding the Left bracket:

- Click or select **Insert / Boss-Base / Extrude**.

- End Condition: **Blind**   - Depth: **.500 in**.

- Click **OK**.

Sketch
Face

20. Creating the 2nd Mounting bracket:

- Select the <u>Face</u> as noted and open a new sketch .

- Either copy the previous sketch – or – recreate the same sketch again on the right side.

- Add Dimensions and Relations needed to fully define the sketch.

Sketch a circle
and 2 centerlines

21. Extruding the Right bracket:

- Click or select **Insert / Boss-Base / Extrude**.

- End Condition: **Blind**.   - Depth: **.500 in**.

- Click **OK**.

22. Constructing the Upper Ring:

- Select the <u>Face</u> as indicated and open a new sketch.

- Sketch the profile below; use Convert Entities where needed.

- The centers of the Ø.250 Circles are coincident with the Ø7.375 bolt-circle.

Sketch Face

Ø7.375
(Bolt Circle)

Convert Entity

Ø.250

Ø8.000

60.00°

Sketch circle

23. Extruding the Upper Ring:

- Click ⬛ or select **Insert / Boss-Base / Extrude**.

- End Condition: **Blind**.   - Depth: **.400 in**. (upward)

- Click **OK**.

24. Adding a Seal-Ring bore:

- Select the <u>Face</u> as noted and open a new sketch ⬛ .

- Select the **inside circular edge** and click **Offset Entities** ⬛ .

- Enter **.150 in**. for Offset Value (larger diameter).

- Click **OK**.

Sketch Face

Offset **.150 in**. from the inside circular edge.

25. Extruding Cut the Seal Ring bore:

- Click or select **Insert / Cut / Extrude**.

- End Condition: **Blind**.

- Depth: **.250 in**.

- Click **OK**.

26. Adding fillets:

- Click or select **Insert / Features / Fillet-Round**.

- Enter **.060 in**. for radius value.

- Select the **7 edges** as noted.

- Click **OK**.

Edges
to fillet

27. Adding Chamfers:

- Click or select **Insert / Features / Chamfer**.

- Enter **.060 in**. for Depth.

- Enter **45 deg**. for Angle.

- Select the **edges** of the 6 holes.

- Click **OK**.

Select the front & back Edges of the holes, on both sides.

28. Saving your work:

- Select **File / Save As / Water Meter Housing / Save**.

Questions for Review

Loft Vs. Sweep

1. In a new part mode, when the Sketch Pencil is selected first, SOLIDWORKS will prompt you to select a sketch plane.
    a. True
    b. False

2. It is sufficient to create a Parallel-Plane-At-Point with a Reference Plane and a Reference Point.
    a. True
    b. False

3. A loft feature uses multiple sketch profiles to define its shape.
    a. True
    b. False

4. Only one guide curve can be used in each loft feature.
    a. True
    b. False

5. Multiple guide curves can be used to connect and control the loft feature.
    a. True
    b. False

6. The guide curves can be either a 2D sketch or a 3D curve.
    a. True
    b. False

7. The loft profiles and the guide curves should be related with Coincident or Pierce relations.
    a. True
    b. False

8. The loft profiles should be created before the guide curves.
    a. True
    b. False

7. TRUE	8. TRUE
5. TRUE	6. TRUE
3. TRUE	4. FALSE
1. TRUE	2. TRUE

Exercise: Loft

There are several different ways to model this part, but this exercise focuses on the Loft technique.

1. Create the part below, using the Loft and the Circular Pattern features.

2. Use the instructions on the following pages, if needed.

1. Starting with the base sketch:

- Select the <u>Front</u> plane and open a new sketch.

- Sketch the profile shown.

- Add the dimensions to fully define the sketch.

2. Revolving the base:

- Click or select **Insert / Boss-Base / Revolve**.

- Direction 1: **Blind**.

- Revolve Angle: **360deg**.

- Click **OK**.

3. Creating the 1st loft profile:

- Select the <u>face</u> indicated and open a new sketch.

- Sketch the profile shown by converting the 2 circular edges then add 2 vertical lines.

- Trim the circles to form one continuous profile.

- <u>**Exit**</u> the sketch.

Sketch face

Symmetric

4. Creating the 2nd loft profile:

Sketch face

- Select the <u>upper face</u> as indicated and open a new sketch.

- Using the same techniques as in the previous sketch, construct this new sketch the same way.

- Add the Collinear relations to both sides, between the 2 sketches.

Collinear both sides

- **Exit** the sketch.

5. Creating the 1st loft feature:

- Click or select: **Insert Boss-Base Loft**.

- Select the outer corner of each sketch profile to prevent the loft from twisting.

- Click **OK**.

6. Pattern and save:

- Create a circular pattern of the lofted feature with a total of 4 instances, then save the part as: **Loft_Exe**.

CHAPTER 7

Loft with Guide Curves

Loft with Guide Curves

This lesson demonstrates the creation of a waved washer using the loft technique, where 4 sketch profiles and a single 3D guide curve are used.

A loft feature normally contains several sections, one centerline parameter and one or more guide curves.

In this case study, four identical sketches will be used as the loft profiles and a single guide curve will be used to connect the sections.

Since the loft profiles are identical, the derived-sketch option will be used to show how the sketches can be derived or copied.
A derived sketch is driven by the original sketch. It can only be positioned with relations or dimensions, but its sketched entities cannot be changed.

When the Original sketch is changed, the derived-sketch will be updated automatically; however, the derived sketches can be Un-derived to break their associations with the Original sketch.

The loft profiles are connected with a 3D curve;
the 3D curve will be generated using the command
called: Curve-Through-Reference-Points. The loft
profiles are also Pierced to these reference points.

The Curve-Through-Reference-Points will be used to guide and control the transition between the loft profiles.

Loft with Guide Curves
Waved Washer

Dimensioning Standards: **ANSI**

Units: **INCHES** – 3 Decimals

Tools Needed:

Insert Sketch	Line	Circle
Sketch Point	Add Geometric Relations	Dimension
Derived Sketch	Curve Through Reference Points	Base/Boss Loft

1. Creating the 1st Construction profile:

- Select the <u>Top</u> plane from the FeatureManager tree.

- Click [icon] or select **Insert / Sketch**.

- Sketch a **Circle** [icon] and convert it into a construction circle [For construction] .

- Add **8 sketch points** [icon] approximately as shown.

Horizontal
top & bottom

Vertical

Horizontal

Vertical
left &
right

2. Fully defining the sketch:

- Select the dimension tool [icon] and add dimensions to the sketch points.

- Add a vertical and a horizontal [icon] relation as indicated.

Point 2

45.00°

Point 3

Point 1

Ø2.000

> 💡 **3-Point Angle Dimension**
>
> To create the 45° dimension, select the Smart Dimension tool and click on the 3 points (starting at the origin) as indicated.

- **Exit** the Sketch [icon] .

3. Creating an Offset Distance plane:

- Click or select **Insert / Reference Geometry / Plane**.

- Select the **Top** plane as Reference Entities.

- Select **Offset Distance** option and enter **.150 in**.

- Click **OK**.

4. Creating the 2nd construction profile using Derived Sketch:

- Hold the **Control** key, select the **new Plane** (plane1) and the **Sketch1** from FeatureManager tree.

- Click **Insert / Derived Sketch**.

5. Positioning the Derived Sketch:

- Add a Vertical & Coincident relation ⊥ between the indicated points.

Vertical

Coincident

- **Exit** the Sketch or select **Insert / Sketch**.

6. Creating a Curve through Reference Points:

- Click or select **Insert / Curve / Curve-Through-Reference-Points**.

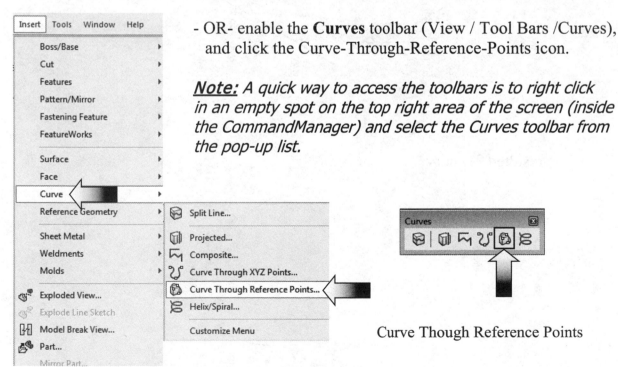

- OR- enable the **Curves** toolbar (View / Tool Bars /Curves), and click the Curve-Through-Reference-Points icon.

Note: A quick way to access the toolbars is to right click in an empty spot on the top right area of the screen (inside the CommandManager) and select the Curves toolbar from the pop-up list.

Curve Though Reference Points

- Select the sketch points in the order as shown (required for this lesson only). (Point 1 is on top, point 2 is on the bottom, 3 on top, 4 on bottom, and so on).

- Click ☑ Closed curve to close the curve.

- Click **OK**.

- The resulted 3D curve.

7. Sketching the 1ˢᵗ loft section:

- Select the <u>Front</u> plane from the FeatureManager tree.

- Sketch a **Rectangle** and add Dimensions and Relations as shown.

Upper vertex is Coincident with a point

.060

.450

- <u>**Exit**</u> the sketch or select **Insert / Sketch**.

8. Creating the 2ⁿᵈ loft section using Derived-Sketch:

- Hold the **Control** key, select the **Front** plane and the sketch of the **rectangle** from the Feature tree.

- Select **Insert / Derived Sketch**.

- The derived sketch is created and placed on top of the original sketch; drag it to the left side and then add a coincident relation to the point on the bottom as noted.

Coincident

9. Fully defining the Derived Sketch:

- Add a **Collinear** relation between the 2 lines as indicated.

Collinear

> ### 💡 Derived Sketch
>
> A derived sketch is a dependent copy of the original sketch. It can only be moved or positioned on the same or different plane with respect to the same model.

- **Exit** the Sketch or select **Insert / Sketch**.

10. Sketching the 3rd loft section: (or use the Derived Sketch option)

- Select the Right plane from the FeatureManager tree.

- Sketch a **Rectangle** (or copy and paste the previous sketch).

- Add Dimensions and Relations needed to fully define the sketch.

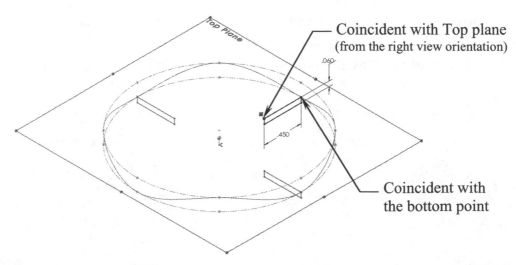

Coincident with Top plane
(from the right view orientation)

Coincident with
the bottom point

- **Exit** the Sketch or select **Insert / Sketch**.

11. Creating the 4th loft section using Derived-Sketch:

- Hold the **Control** key; select the **Right** plane <u>and</u> the sketch of the 3rd **rectangle** from the FeatureManager tree.

- Select **Insert / Derived Sketch**.

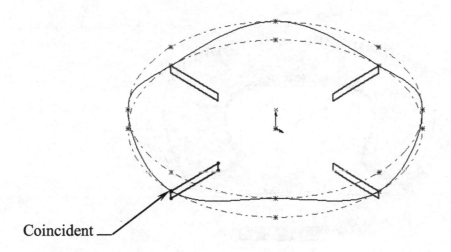

Coincident

12. Constraining the Derived sketch:

- Add a **Collinear** relation between the 2 lines as shown.

Collinear

- <u>**Exit**</u> the Sketch or select **Insert / Sketch**.

13. Creating a Loft with Guide curve:

- Click or select **Insert / Boss-Base / Loft**.

- For Loft Profiles: Select the <u>four</u> rectangular sketches.

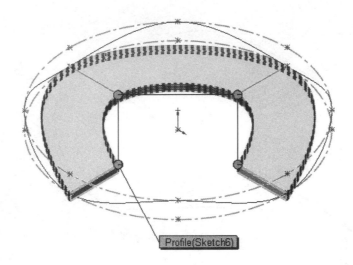

Profile(Sketch6)

- For Guide Curve, select the <u>3D curve</u>.

- Click **Close Loft** ☑ Close loft under Options (arrow).

Guide Curve(Curve1)

Profile(Sketch6)

- Click **OK**.

- The completed Waved Washer (with the construction sketches still visible).

14. Hiding the construction sketches:

- Right click on the construction sketches and select **Hide** .

Hide sketch —

15. Saving your work:

- Select **File / Save As**.

- **Enter Waved Washer** for the file name.

- Click **Save.**

Questions for Review

Loft with Guide Curves

1. A sketch profile can be copied onto another plane or a planar surface.
   a. True
   b. False

2. Sketch points can be added in any sketch to help define the sketch geometry or locations.
   a. True
   b. False

3. If a derived sketch is driven by the original sketch, its entities cannot be changed.
   a. True
   b. False

4. A 3D curve can be created using the reference points in the sketches or model's vertices.
   a. True
   b. False

5. The loft sections should either be Pierced or Coincident with the guide curves.
   a. True
   b. False

6. The guide curves can also be used to control the loft sections from twisting.
   a. True
   b. False

7. Only two guide curves can be used in each loft feature.
   a. True
   b. False

8. Up to four sketch profiles can be used in a loft feature.
   a. True
   b. False

9. The construction sketches can be toggled (Show/Hide) at any time.
   a. True
   b. False

9. TRUE
7. FALSE   8. FALSE
5. TRUE   6. TRUE
3. TRUE   4. TRUE
1. TRUE   2. TRUE

Exercise: Using Curve Driven Pattern

1. Creating the Base sketch:

- From the <u>Top</u> plane sketch a **circle**, centered on the origin.

- Add a diameter dimension of **.100"**.

2. Activating the Surfaces toolbar:

- Right click the **Sketch Tab** and select the **Surfaces** option to enable it.

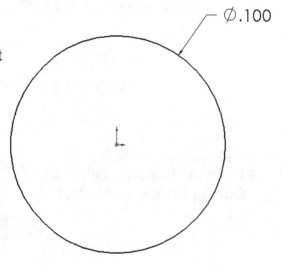

3. Extruding a <u>surface</u>:

- From the Surfaces toolbar click: **Extruded Surface**.

- Use the **Blind** type and enter **.060"** for extrude depth.

- Click **OK**.

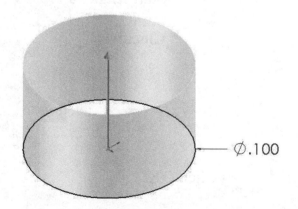

4. Creating a helix:

- Select the <u>Top</u> plane and open a new sketch.

- Select the **bottom edge** of the extruded surface and press **Convert Entities**.

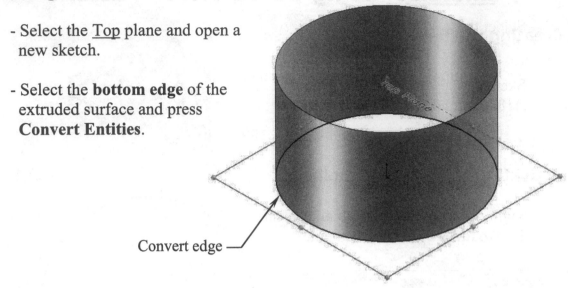

Convert edge ⎯⎯⎯

- From the Features toolbar click: **Curves / Helix and Spiral**.

- Enter the following:

  * Defined by: **Pitch and Revolution**

  * **Constant Pitch**

  * Pitch: **0.020"**

  * Revolutions: **3**

  * Start Angle: **0**

  * **Clockwise**

- Click **OK**.

5. Creating a surface profile:

- Select the <u>Top</u> plane and open a new sketch.

- Sketch a **vertical line** and add a **.010"** linear dimension.

- Add a **Pierce** relation between the endpoint of the line and the helix.

Pierce relation —

.010

- Switch to the **Surfaces** toolbar and click **Extruded Surface**.

- For extrude type, use the default **Blind** type.

- Enter **.020"** for extrude depth.

- Click **OK**.

.010

6. Creating a Curve Driven Pattern:

- From the Features toolbar and click:
  Linear Pattern / Curve Driven Pattern.

- For Direction 1, select the **Helix** (Edge 1).

- Enter **61** for Instances.

- Enable the **Equal Spacing** checkbox.

- Under Curve Method select:
  Transform Curve.

- For Alignment Method, select:
  Tangent to Curve.

- Under Face Normal select the **Cylindrical surface** of the cylinder.

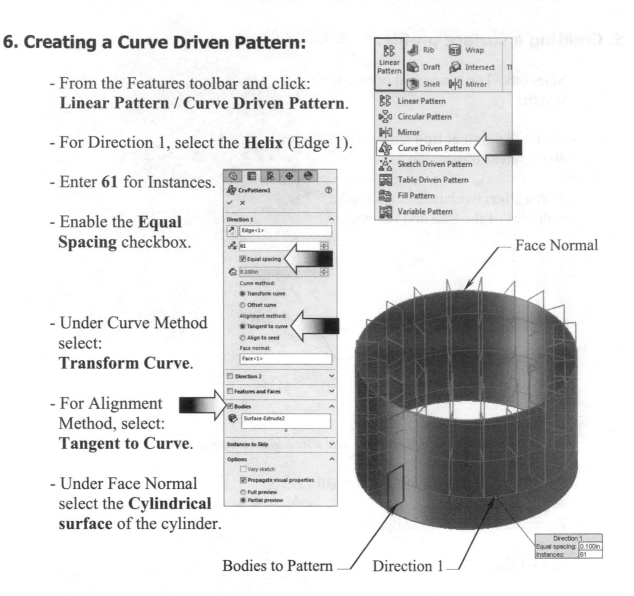

Face Normal

Bodies to Pattern —   Direction 1 —

Direction 1	
Equal spacing:	0.100in
instances:	61

- Expand the **Bodies to Pattern** section and select the **Extruded Surface** in step 5.

- Click **OK**.

Hide this surface

7. Creating a Curve Through Reference Points:

- From the Features toolbar select:
Curves / Curve Through Reference points.

- Click the starting point (point 1) at the end
of the helix and go clockwise to point 2,
then point 3 as indicated.

Point 2 ——⌐   ⌐—— Point 1

- Continue going around and select all connecting points. Simply delete any
mistakes from the Through Points dialog box.

- Be sure to **uncheck** the Close Curve checkbox since the two ends of this curve
are not supposed to connect.

- Click **OK**.

——Hide all surfaces of the Curve Driven Pattern

Also hide —
the helix

8. Creating the final sweep:

- Switch to the Features toolbar and click: **Swept Boss Base**.

- Select the **Circular Profile** option (arrow).

- Enter **.002in** for Profile Diameter (arrow).

- Select the **Curve1** as the sweep path.

- Click **OK**.

9. Saving your work:

- Click **File / Save As**.

- Enter **V-Shape Spring** for the name of the file.

- Click **Save**.

- Close all documents.

Advanced Sweep

Advanced Sweep
Wire Form

This second half of the chapter discusses one of the advanced techniques on creating a continuous wrapped wire around a coil.

There will be two separate sweep features created in this design. The first sweep feature is the coil, shaped like a spring; the sweep profile is made tangent to the sweep path. To achieve this, a couple of construction lines are used to help locating one of its quadrant points on the path.

The second sweep feature, due to its complex shape, is done by using a 3D sketch as a sweep path. It is going to wrap around the first sweep feature in a unique way that makes 3D sketch one of the few options to create it with.

The method in this lesson demonstrates the use of locating points created in two different sketches to guide a 3D sketch. The lines in the 3D sketch are connected to the pre-defined endpoints of the lines, and the shape of the wire, or the sweep path is formed.

To ensure the point locations are identical between the two sketches, the option Derived-Sketch is used. That way if one of the endpoints in the original sketch is moved, the same point in the derived sketch will also move.

The entities in the derived sketch are driven by the original sketch. Changes done to the original sketch are reflected in the derived sketch. To break the link between the derived sketch and its parent sketch, right click the derived sketch from the FeatureManager tree and select **Underived**. After the link is broken, the derived sketch will no longer update when the original sketch is changed.

Advanced Sweep
Wire Form

Dimensioning Standards: **ANSI**

Units: **INCHES** – 3 Decimals

Tools Needed:

Insert Sketch	3D Sketch	Line
Add Geometric Relations	Sketch Fillet	Circle
Dimension	Centerline	Fillet/Round
Sketch Circular Pattern	Helix/Spiral	Sweep Boss/Base

1. Starting a new part document:

- Go to **File / New / Part**.

- Set the Drafting Standard to **ANSI** and the Units to **Inches, 3 decimals**.

- Select the <u>Top</u> plane and open a new sketch 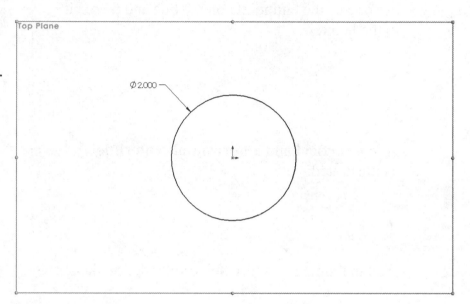.

- Sketch a **Circle** centered on the origin as shown.

- Add a diameter of **Ø2.000"** to fully define the sketch.

2. Creating the sweep path:

- Click the 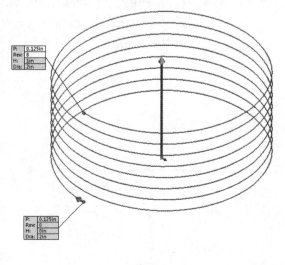 **Helix / Spiral** command on the Curves drop down menu, or select: **Insert / Curve Helix-Spiral**.

- Set the following:

    * **Constant Pitch**
    * Pitch = **.125"**
    * Revolutions: **8**
    * Start Angle = **0deg**
    * **Clockwise**

- Click **OK**.

3. Creating the sweep profile:

- Select the <u>Right</u> plane and open a new sketch .

- Sketch a **Circle** on this plane.
  The lower quadrant point of the circle must
  be pierced to the helix, and one way to achieve
  that is to create a couple of centerlines and then
  use one of the endpoints on the line and pierce it
  to the helix.

- Add a **vertical** and a **horizontal centerline**
  as illustrated.

Ø.120

- Add the dimension **Ø.120"** to fully define the sketch.

Add a Vertical and a
Horizontal centerline

- Add a **Pierce** relation between the
  endpoint of the vertical centerline
  and the helix.

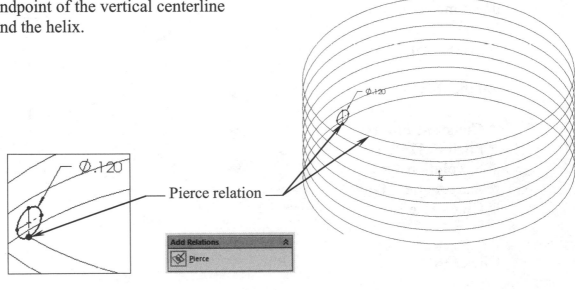

Ø.120

Pierce relation

Add Relations	
Pierce	

4. Creating the sweep feature:

- Switch to the **Features** tool tab.

- Click the **Swept Boss / Base** command.

- For Sweep Path, select the **Helix**.

- For Sweep Profile, select the small **Circle**.

- Leave all other parameters at their default settings.

- Click **OK**.

5. Creating the 1st offset plane:

- From the Features tool tab, click **Reference Geometry / Plane** or select: **Insert / Reference Geometry / Plane**.

- For the First Reference select the **Top** plane from the Feature-Manager tree.

- Under Offset Distance enter: **1.150"**.

- Place the new plane <u>above</u> the Top reference plane.

- Click **OK**.

6. Creating the 2nd offset plane:

- Click the **Plane** command once again.

- Select the **Top** plane for First Reference.

- Enter **.045"** for Offset Distance.

- Place the new plane <u>under</u> the Top reference plane.

- Click **OK**.

7. Sketching on the 1st offset plane:

- Select the **Plane1** and open a new sketch .

- Sketch **2 Circles** that are centered on the origin.

- Sketch a vertical **Line** and ensure that both ends of the line are coincident with the upper quadrant points of the two circles.

- Add the two diameter dimensions shown.

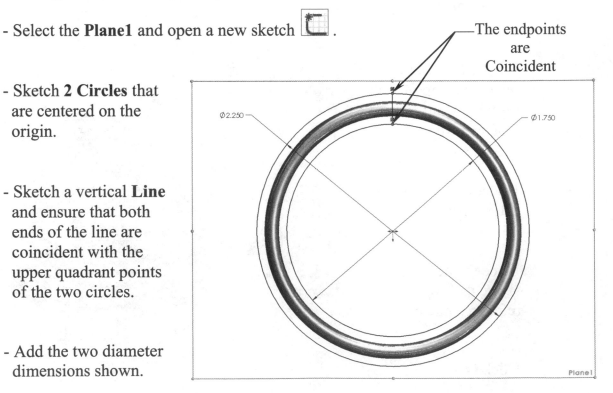

The endpoints are Coincident

Ø2.250

Ø1.750

Plane1

- Hold the **Control** key and select both circles.

- Release the control key and enable the **For Construction** checkbox on the Property tree.

- This option toggles the selected entities back and forth between a line and a construction line.

Convert to construction lines

8. Creating a Circular Sketch pattern:

- Select the vertical line (Entity to Pattern) and click the **Circular Sketch Pattern** command.

Entity to pattern

- The origin is automatically selected as the center of the pattern.

- Enable the **Equal-Spacing** checkbox.

- Enter **32** for the number of instances.

- Click **OK**.

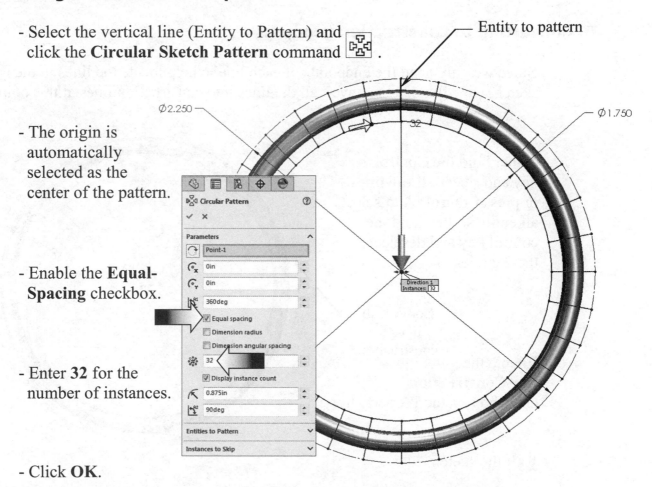

- The pattern is still under defined at this point.

- Hold the **Control** key and select the 3 points as noted.

- Select the **Horizontal** relation from the Property tree.

- The sketch turns black to indicate a Fully Defined status. Click **OK** to exit the Property tree.

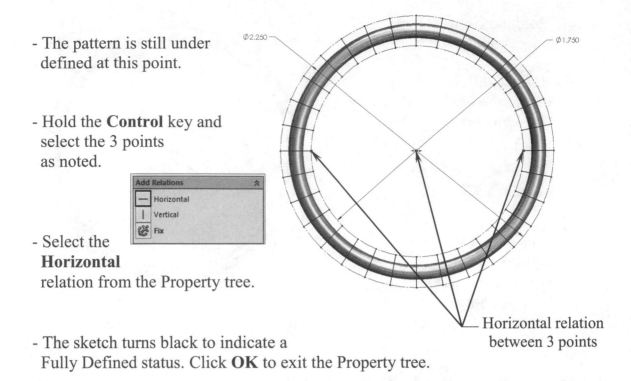

Horizontal relation between 3 points

9. Converting to construction geometry:

- Since we only need the endpoints of each line to help locate the lines in the next sketch, we will need to convert all the lines into construction lines at this point.

- Either hold the Control key and select **all the lines**, or press Control+A to select all entities, then hold the control key and deselect the 2 circles.

Convert all lines to construction

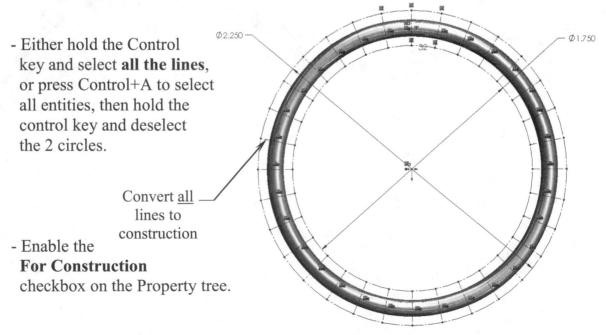

- Enable the **For Construction** checkbox on the Property tree.

- **Exit** the sketch.

10. Creating a derived sketch:

- Hold the **Control** key and select the **Plane2** and the **Sketch3** from the feature tree.

- Click **Insert / Derived Sketch** (arrow).

- A copy of the sketch is created on the selected plane.

Coincident relation

- Add a **Coincident** relation between the 2 center points as noted.

- Add a **Collinear** relation between the <u>horizontal line</u> on the right and the <u>Front plane</u> as indicated.

- All entities in the sketch should change to the black color at this point.

- **Exit** the sketch.

Collinear relation

11. Creating a 3D sketch:

- The endpoints on the lines will help us in creating the 3D sketch a little easier and also more accurately. Make sure both of the sketches are visible.

- From the Sketch tool tab, click the drop-down arrow and select **3D Sketch** 3D .

- Select the **Line** tool from the 3D Sketch toolbar.

- Sketch a line that starts at one of the endpoints on top and connect it to one of the endpoints down below.

- The feature Swept1 is hidden for clarity.

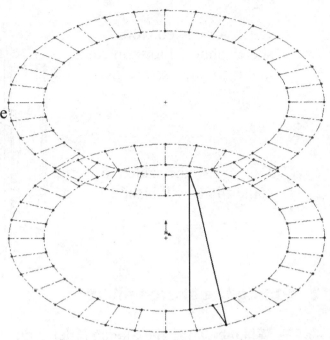

(Use the Click + Release technique to create multiple lines, while the Click +Hold + Drag creates only 1 line each time.)

- Add 3 more lines as shown. Be sure to snap to each existing endpoint when sketching the lines.

- Continue to add the other lines as shown below. If you run into an error, simply press Escape and start sketching the lines again.

- It is maybe a little easier if you could change the orientation of the view after a few lines.

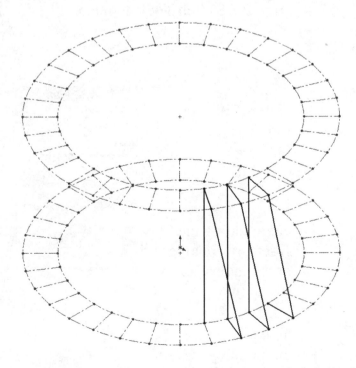

- While in the middle of the line mode you can press and hold the scroll-wheel to rotate slightly to a different angle, then continue with your sketch.

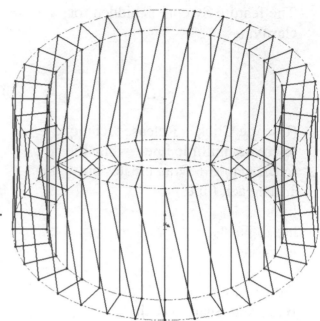

- Once completed, your sketch should look something like the image shown here.

- Do not exit the sketch just yet; some fillets will be added next.

12. Adding the sketch fillets:

- When creating the Sketch Fillets you can either select the 2 lines, or click directly at the intersection point between the 2 lines to add the fillets. If a fillet gets deleted its sharp corner will reappear.

- Click the **Sketch Fillet** command from the Sketch Tools tool tab.

- Enter **.120"** for radius value.

- Start adding the fillets by clicking at the intersections of the lines.

- Continue with adding the same fillet
  to the other lines.

- It does not really matter which direction
  (clockwise or counter clockwise) when
  adding the fillets, but it does make it a
  little easier to go around 1 direction and
  not to jump back and forth as you may
  miss one or more vertices.

- After completed, your sketch should look like the image shown below. This 3D
  sketch will be used as the sweep path in the next few steps.

- **Exit** the 3D sketch.

13. Creating the sweep feature:

- Switch to the **Features** tool tab.

- Select the **Swept Boss/Base** command .

- Select the Circular Profile option (arrow).

- Enter **.080in** for Profile Diameter (arrow).

- For Sweep Path, select the **3D Sketch**.

- Click **OK**.

14. Saving your work:

- Click **File / Save As**.

- For the file name, enter: **Advanced Sweep_Wire Form**.

- Select a location to save the file and click **Save**.

Exercise: Using Curve Through Reference Points

1. Creating the Base sketch:

- From the <u>Top</u> plane sketch a circle and a horizontal Centerline, then convert both entities to Construction lines.

- Add a diameter dimension of **.100"**.

- Select the Horizontal Centerline and click **Circular Sketch Pattern**.

- Enter **16** instances, **Equal Spacing**.

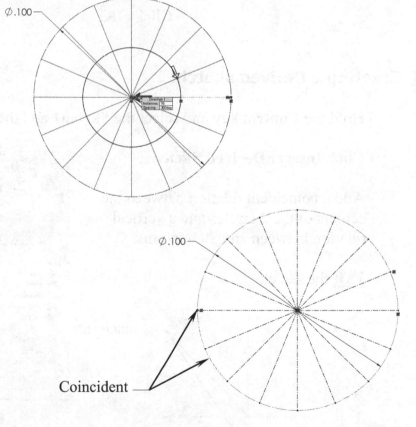

Coincident

- Add a couple of Coincident relations between the endpoints of any of the instances and the circle, to fully define the sketch.

- <u>**Exit**</u> the sketch (or press **Control +Q**).

2. Creating an offset plane:

- Create a new plane that is **.020"** <u>above</u> the **Top** plane.

- Click **OK**.

3. Creating a Derived sketch:

- Hold the **Control** key and select the **Sketch1** and the **Plane1**.

- Click **Insert / Derived Sketch**.

- Add a coincident relation between the centers of thc 2 circles and a vertical relation between any 2 endpoints.

- <u>**Exit**</u> the sketch.

Coincident

Vertical

4. Creating a Curve Through Reference Points:

- From the **Features** toolbar, click:
Curves / Curve Through Reference Points.

- Click the starting point (point 1), move
clockwise and click point 2, then point 3
and so on. Continue around.

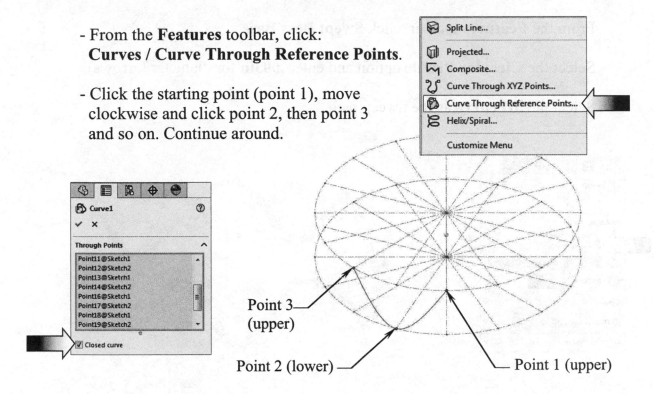

Point 3 (upper)

Point 2 (lower)

Point 1 (upper)

- Continue going around a full revolution and select all the connecting points in
the 2 sketches.

- At the end of the path, do not click point 1 again as the **Closed Curve** option
will join the 2 ends together automatically.

- Enable the **Closed Curve** checkbox.

- Click **OK**.

Stop here

5. Creating a swept feature:

- From the **Features** toolbar, click **Swept Boss Base**.

- Select the **Circular Profile** option and enter **.003in** for diameter (arrows).

- Select the **3D curve** as the Sweep path.

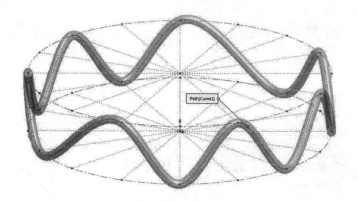

- Click **OK**.

- Hide the construction sketches and the 3D curve.

6. Saving your work:

- Click **File / Save as**:

- Enter: **Curve Through Reference Points**
  for the name of the file.

- Click **Save**.

CHAPTER 8

Using Surfaces

Using Surfaces
Advanced Modeling

Surfaces are a type of geometry that can be used to create solid features.

The surface options are used to form complex free-form shapes and to manipulate files imported from other Cad formats.

Unlike solid models, surfaces can be opened, overlapped, and have no thickness. Each surface can be constructed individually and then knitted together. A solid feature is created by thickening the surfaces that have been knitted into a closed volume.

Surfaces can be modeled in any shape and their sketches can either be extruded, revolved, swept, or lofted into a surface. These surfaces can also be replaced and filled with other surfaces.

Edges of the surfaces can be extended and trimmed. Surfaces can be moved, rotated, and copied.

The angle between the faces of a surface can be calculated using the Draft-Analysis tool; Positive Drafts, Negative Drafts, and Required Drafts are reported on screen.

This first half of the chapter discusses the use of some surfacing tools in the newer releases of SOLIDWORKS.

Using Surfaces
Advanced Modeling

Dimensioning Standards: **ANSI**

Units: **INCHES** – 3 Decimals

Tools Needed:

Insert Sketch	3 Point Arc	Ellipse
Dimension	Add Geometric Relations	Split Line
Plane	Lofted Surface	Surface Thicken

1. Constructing a new work plane:

- Select the <u>Front</u> plane from the FeatureManager tree.

- Click or select **Insert / Reference Geometry / Plane**.

- Select the **Front** plane and click the **Offset Distance** button.

- Enter **3.000 in**. and enable the **Flip Offset** check box to reverse the direction.

- Click **OK**.

2. Sketching the 1ˢᵗ profile:

- Select the <u>new plane</u> (Plane1).

- Click or select **Insert / Sketch**.

- Sketch a **3-Point-Arc** and add the dimensions as shown.

- Add a **Vertical** and **Horizontal** relations as noted. The center of the arc is .625 below the origin.

- <u>**Exit**</u> the Sketch .

3. Sketching the 2ⁿᵈ profile:

- Select the <u>Front</u> plane from the FeatureManager tree.

- Click 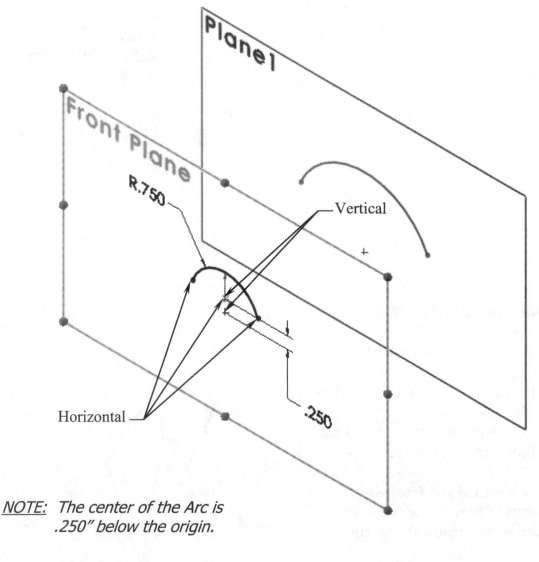 or select **Insert / Sketch**.

- Sketch a **3-Point-Arc** as shown.

- Add a **.750** radius dimension to the arc.

- Add a Horizontal and a Vertical relation as indicated.

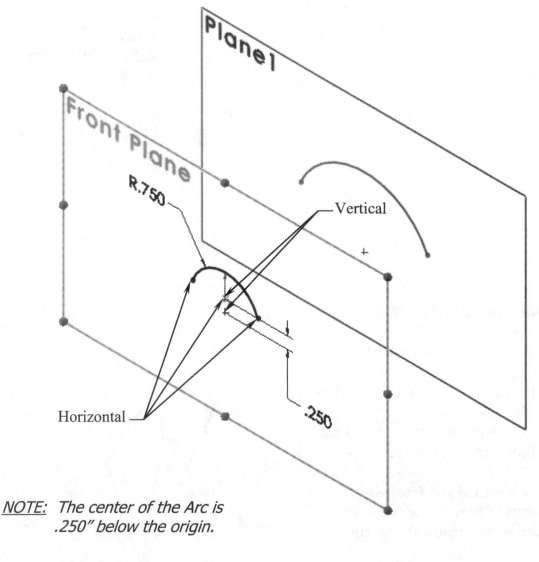

NOTE: *The center of the Arc is*
.250" below the origin.

- **Exit** the sketch .

4. Sketching the Guide Curve:

- Select the <u>Right</u> plane from the FeatureManager tree and open a new sketch.

- Sketch a **3-Point-Arc** and add a **4.00** in. dimension .

- Add the **Pierce** relations as noted .

- **Exit** the sketch .

Enabling the Surfaces toolbar: *Right-click on one of the tabs (Features, Sketch, etc.) and enable the Surfaces option. The new Surfaces tab appears next to the other tabs.*

5. Creating a Surface-Loft:

- Click 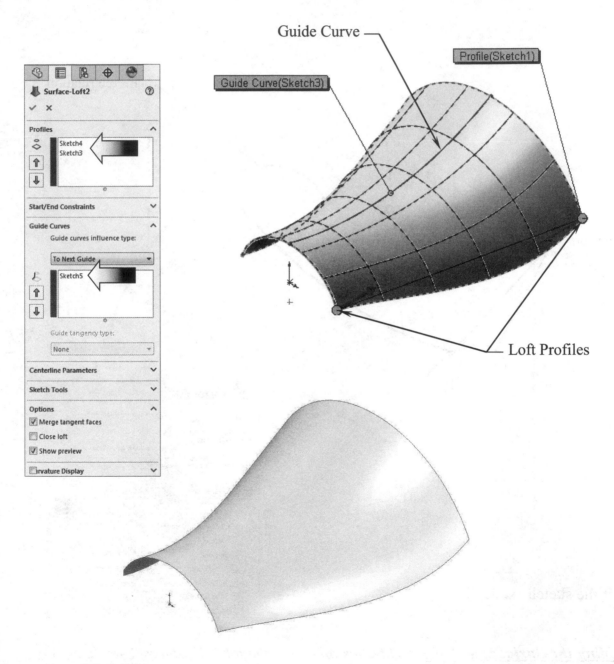 on the Surfaces tool tab or select: **Insert / Surface / Loft**.

- Select the **2 Sketched Profiles** by clicking on their *right-most endpoints*.

- Expand the Guide-Curve section and select the **Sketched Arc** as noted.

- Click **OK**.

6. Sketching the Split profile:

- Select the Top plane from the FeatureManager tree.

- Click or **Insert / Sketch**.

- Change to Top view orientation .

- Sketch an **Ellipse** and add Dimensions/Relations as shown.

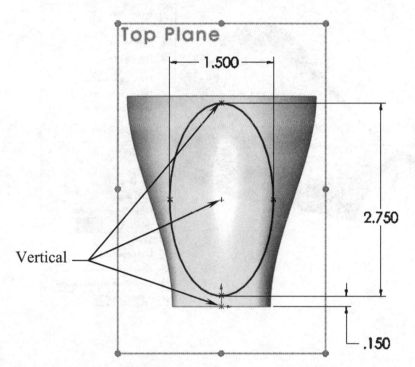

Vertical

> ### 💡 Split Lines
>
> The Split Lines command projects a sketch entity onto a face (or a set of faces) and divides a selected face into multiple separate faces, enabling you to select and work with each face.

7. Splitting the surface:

- Click on the Curves toolbar OR select **Insert / Curve / Split Line**.

- Select the **Ellipse** as Sketch-to-Project .

- Select the **Surface-Loft1** as Faces-to-Split .

- Click **OK**.

8. Deleting Surfaces:

- Right click on the **outer portion** of the surface.

- Select **Face** / **Delete** from the menu.

- Click **Delete** under options (circled).

- Click **OK**.

💡 **Delete Face**

This command deletes one
or more faces from a surface
or solid body.

Other options are:
* Delete and Patch, which
  automatically patches and
  trims the body.
* Delete and Fill, which
  generates and fills any gap.

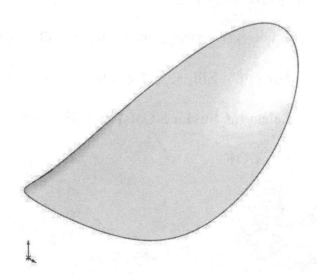

- The resulting surface.

9. Thickening the surface:

- Click 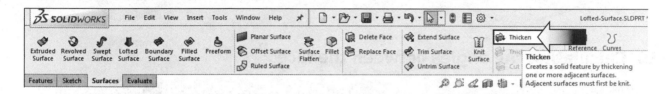 or select **Insert / Boss-Base / Thicken**.

- Click on the **surface** as Surface-To-Thicken.

- Choose **Thicken Both Sides** option.

- Enter **.030** in. as Thickness (.060 total thickness).

Both Sides

> ### 💡 Thicken Surfaces
>
> In order to create a solid volume, all surfaces have to form a closed shape.
>
> If the shape is open, a wall thickness can be added to the surface to close.

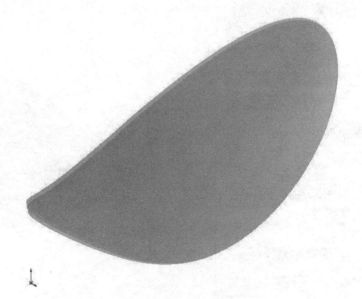

- Click **OK**.

- The surface model is thickened into a solid model.

10. Calculating the angles between the faces:

- Change to the **Evaluate** tool tab and click the **Draft Analysis** command.

- Select the **Top** plane for Direction of Pull.

- Enter **1.00** deg. for Draft Angle.

- Enable the **Face Classification** checkbox and click **Calculate**.

- The **light blue color** indicates the surfaces that have both positive and negative drafts on them. This can be eliminated by creating a split line in the middle, or by adding a full round fillet around the parameter.

> ☼ **Draft Analysis**
>
> Using the settings in draft analysis, you can verify the draft angles on model faces or you can examine angle changes within a face.

Draft Analysis

The **Draft Analysis** is a tool to check the correct application of draft to the faces of each part. With draft analysis, you can verify draft angles, examine angle changes within a face, as well as locate parting lines, injection, and ejection surfaces in parts.

Draft analysis results listed under Color Settings are grouped into four categories, when you specify Face classification:

Positive draft: Displays any faces with a positive draft based on the reference draft angle you specified. A positive draft means the angle of the face, with respect to the direction of pull, is more than the reference angle.

Negative draft: Displays any faces with a negative draft based on the reference draft angle you specified. A negative draft means the angle of the face, with respect to the direction of the pull, is less than the negative reference angle.

Draft required: Displays any faces that require correction. These are faces with an angle greater than the negative reference angle and less than the positive reference angle.

Straddle faces: Displays any faces that contain both positive and negative types of draft. Typically, these are faces that require you to create a split line.

*Note: When analyzing the draft for surfaces, an additional **Face classification** criterion is added: **Surface faces with draft**. Since a surface includes an inside and an outside face, surface faces are not added to the numerical part of the classification (**Positive draft** and **Negative draft**). Surface faces with draft lists all positive and negative surfaces that include draft.*

- Hover the pointer over the upper surface to see the read out draft angle for that particular area.

- Position the mouse cursor over the yellow areas (required drafts) and check the draft angles to see if they meet your draft requirements.

- Click **OK**.

- The option Deviation-
Analysis can be used to
diagnose and calculate
the angle between faces.

- The colored arrows
display the amount of deviation.
The results show the Max,
Min and Average deviations
between the adjacent faces.

11. Adding a Full Round Fillet:

- Click or select **Insert / Features / Fillet-Round**.

- Select the Side-Face-Set1, Center-Face-Set, and Side-Face-Set2 as noted.

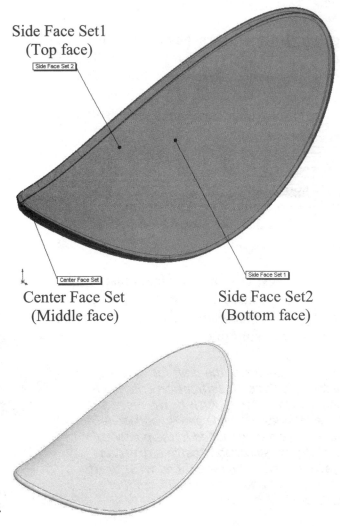

Side Face Set1
(Top face)

Center Face Set
(Middle face)

Side Face Set2
(Bottom face)

- Click **OK**.

- The resulting full round fillet.

12. Creating an Offset Distance plane:

- Select the **Top** plane from the FeatureManager tree and click , or select **Insert / Reference Geometry / Plane**.

- Enter **.850 in**. for Distance and place the new plane <u>above</u> the Top plane.

- Click **OK**.

13. Sketching the Slot Contours:

- Select the new plane (**Plane2**) and open a new sketch 📄.

- Create the sketch using either **Mirror** or **Offset** options.

- Sketch the profile as shown; add dimensions to fully position the sketch.

- Create an offset 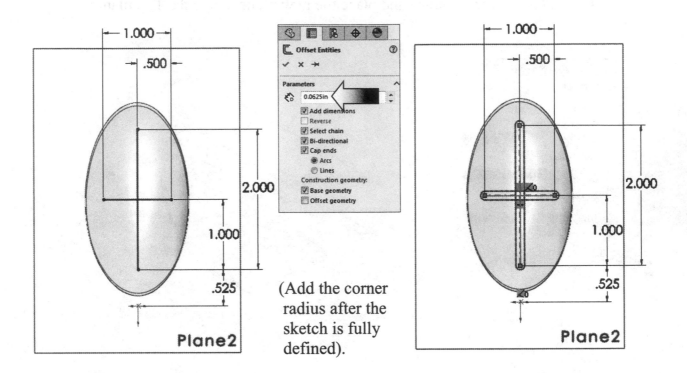 of **.0625 in.** from the 2 sketch lines, using the settings below.

(Add the corner radius after the sketch is fully defined).

- Trim the inner intersections and create a second offset also at .0625in as shown.

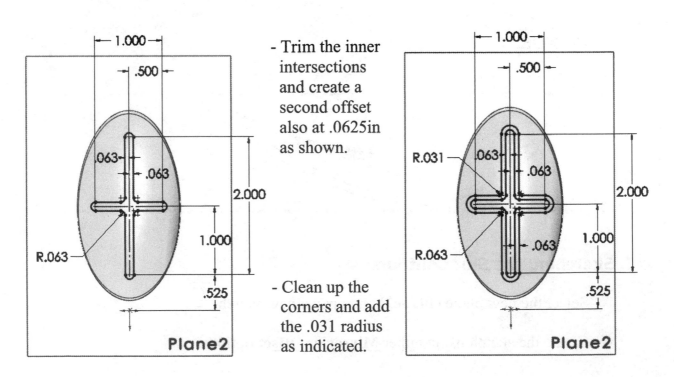

- Clean up the corners and add the .031 radius as indicated.

14. Extruding Cut the 1st Contour:

- Click or select **Insert / Cut / Extrude**.

- Expand the Selected Contour section and select one of the **Outer Lines** (Outer Contour)

- Use **Offset From Surface** end condition.

- Enter **.030 in**. for Depth.

- Click the **Top face** of the model.

- Enable **Reverse Offset** check box.

- The slot is cut, following the same contours of the upper surface.

- Click **OK**.

Select one of the outer lines (contour)

Offset from this face

15. Extruding Cut the 2nd Contour:

- Expand the Cut-Extrude1 from the FeatureManager tree, right click on the Sketch5 and select **Show** .

- Hover over one of the <u>inner lines</u> and select the entire contour when it highlights.

- Click or select **Insert / Cut / Extrude**.

- Select **Through All** for end condition.

- Click **OK**.

Select all inner entities

- **Hide** the Sketch5 when finished.

16. Saving your work:

- Select **File / Save As / Lofted-Surface / Save**.

Questions for Review

Using Surfaces

1. Surfaces are a type of geometry that can be used to create complex shapes.
    a. True
    b. False

2. Surfaces can be opened, overlapped, and have no thickness.
    a. True
    b. False

3. Surfaces can be modeled into any shape and can be extruded, revolved, swept, or lofted.
    a. True
    b. False

4. The split line option can be used to "divide" a surface into two or more surfaces.
    a. True
    b. False

5. Several surfaces can be lofted together to form a solid feature.
    a. True
    b. False

6. Surfaces cannot be moved or copied in a part document.
    a. True
    b. False

7. Each surface can be created individually and then knitted together as one surface.
    a. True
    b. False

8. The same Sketched profile can be re-used to create different extruded contours.
    a. True
    b. False

9. Offset From Surface (extrude option) only works with surfaces, not solid features.
    a. True
    b. False

9. FALSE
7. TRUE   8. TRUE
5. TRUE   6. FALSE
3. TRUE   4. TRUE
1. TRUE   2. TRUE

Lofted Surface

Let's take a look at a couple of techniques when modeling advanced shapes.

We will start out by creating some surfaces using various surfacing tools. These surfaces will get knitted into one surface; this surface will then get thickened into a solid part. Finally, the part is split into two halves and gets assembled in an assembly document.

1. Creating new offset planes:

- Select the <u>Front</u> plane from the FeatureManager tree.

- Click or select **Insert / Reference Geometry / Plane**.

- Choose **Offset Distance** and enter **4.00 in**.

- Use **Flip** direction if needed to place the new plane on the *back side*.

- Enter **2** for number of instances.

- Click **OK**.

2. Sketch the first profile: (the front section)

- Select the <u>Front</u> plane from the FeatureManager tree.

- Click or select **Insert / Sketch**.

- Sketch the profile and add the dimensions shown.

- **Exit** the Sketch or select **Insert / Sketch**.

3. Sketching the second profile: (the middle section)

- Select the <u>Plane1</u> from FeatureManager tree.

- Click or select **Insert / Sketch**.

- Sketch the profile and add the dimensions shown.

💡 **Copy & Paste**

The 1st profile can be copied and pasted to make the next 2 sketches. The dimensions will then be adjusted to size.

- **Exit** the Sketch or select **Insert / Sketch**.

4. Sketching the third profile: (the end section)

- Select the <u>Plane2</u> from the FeatureManager tree.

- Click or select **Insert / Sketch**.

- Sketch the profile and add dimensions as shown.

- **Exit** the Sketch ⬚ or select **Insert / Sketch**.

5. Selecting the loft profiles:

- To prevent the loft feature from being twisted, it is recommended that all sketch profiles should be selected from the same side.

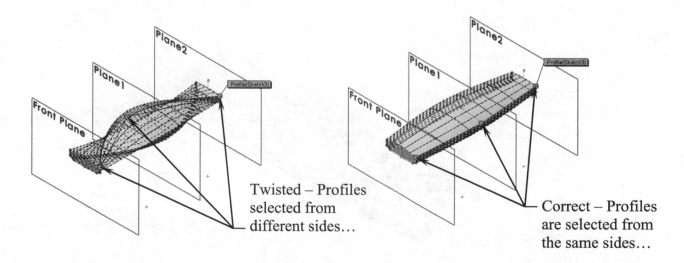

Twisted – Profiles selected from different sides…

Correct – Profiles are selected from the same sides…

6. Lofting between the profiles:

- Click or select **Insert / Surface / Loft**.

- Select the **upper-right vertex** of each profile.

- Enable **Merge Tangent Faces**.

- Click **OK**.

> ### ☀ Merge Tangent Faces
>
> Select Merge tangent faces to cause the corresponding surfaces in the resulting loft to be tangent if the corresponding lofting segments are tangent.

Select the upper vertex of each profile

- The resulting Surface-Loft.

7. Creating a Revolved sketch:

- Select the <u>Top</u> plane from the FeatureManager tree.

- Click ⬜ or select **Insert / Sketch**.

- Sketch the revolve profile and add dimensions as shown.

R4.000

Coincident

4.750

8. Revolving the Spherical surface:

- Click 🌀 or select **Insert / Surface / Revolve**.

> 💡 **Revolved Surface**
>
> Revolved Surface creates a surface by rotating a sketch profile around a centerline (or the Axis of Revolution).

- Revolve Type: **Blind**.

- Revolve Angle: **360 deg**.

- Click **OK**.

- The Revolved Surface.

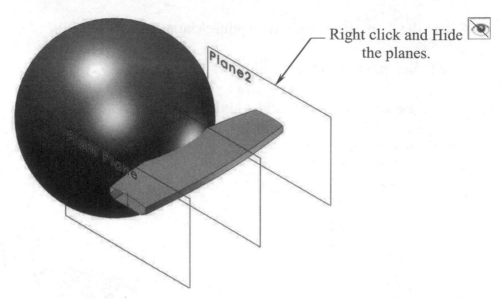

Right click and Hide
the planes.

9. Copying the Revolved Surface:

- Click 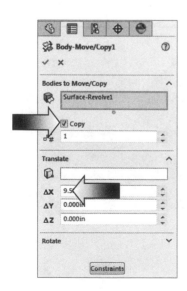 or select **Insert / Surface / Move/Copy**.

- Select the **Surface-Revolve1** under Surfaces to Move/Copy.

- Enable the **Copy** check box.

- Enter **1** for Number of Copies.

- Enter **9.500** in the **Delta X** distance box.

- Click **OK**.

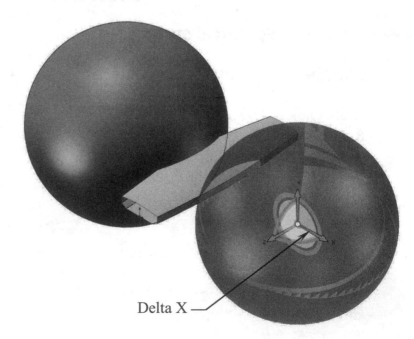

Delta X

10. Trimming the Base part:

- Click or select **Insert / Surface / Trim**.

- For Trim-Type, click **Mutual Trim** `⊙ Mutual trim` .

- For **Trimming-Surfaces**, select all **3 surfaces**.

- For **Keep Selection**, select the Surface-Loft1.

- Click **OK**.

> 💡 **Trim Surface**
>
> A surface or a sketch can be used as a trim tool to trim the intersecting surfaces.

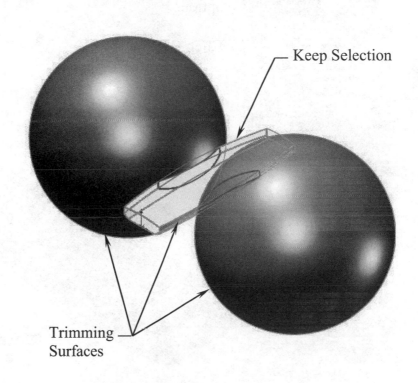

Keep Selection

Trimming Surfaces

11. Hiding the surfaces:

- Right click over the two revolved surfaces and select **Hide** 👁.

12. Filling the openings with Surface-Fill:

- For Patch Boundary: select the **6 edges** as shown.

- For Curvature Control use: **Contact** [Contact ▾].

- Enable **Apply to all edges**.

- Clear the **Optimize Surface** option.

- Click **OK**.

> 💡 **Filled Surface**
>
> Filled surface constructs a surface patch to fill a non-planar opening in a model.

Select 6 edges

Contact - S0 - Boundary

- The right side cutout is filled with a new surface (Surface-Fill1).

13. Filling the left side cutout:

- Rotate the view to the other side or Hold the Shift key and press the Up arrow key twice (rotate 90° per key stroke).

- Repeat step 13 to fill the left side cutout with a new surface.

Select 6 edges

- The left side cutout is filled with a new surface (Surface-Fill2)

14. Filling the openings with Planar Surface:

- Click or select **Insert / Surface / Planar**.

- For Boundary Entities: select all **12 edges** in the front and back openings.

- Click **OK**.

Select 6 edges
in the front

Planar Surface

Planar surface constructs
a surface patch to fill
a planar opening in a
model.

Select 6 edges
in the back

- The front and back openings are filled with planar surfaces (Surface-Plane2).

15. Creating a Surface-Knit:

- Click or select **Insert / Surface / Knit**.

- For Surfaces/Faces-To-Knit, select all **5 surfaces**.

- Click **OK**.

> 💡 **Knit Surface**
>
> Combines two or more faces and surfaces into one.

Select all surfaces

- All 5 surfaces are knitted and combined into one (Surface-Knit1).

16. Adding .500" fillets:

- Click or select **Insert / Features / Fillet/Round**.

- Type **.500 in**. for Radius.

- Select **8 edges** as shown.

- Click **OK**.

Select 8 edges
(on both sides)

Radius: 0.5in

17. Adding .125" fillets:

- Click or select **Insert / Features / Fillet/ Round**.

Radius: 0.125in

Box-Select to
group all edges
(on both sides)

- Enter **.125 in**. for Radius and select **all edges** shown.

- Click **OK**.

- Compare your model with this image. Rotate the model to make sure that all edges are filleted.

18. Creating a solid from the surface model:

- Click 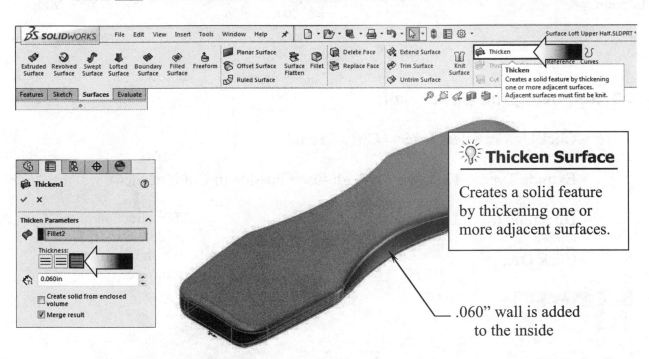 on the Surfaces tool tab or select **Insert / Boss-Base Thicken**.

Thicken Surface

Creates a solid feature by thickening one or more adjacent surfaces.

.060" wall is added to the inside

- For **Surface-To-Thicken**: Select the model from the graphics area.

- For thickness Direction: Select **Thicken Side 2** ▤ (Inside).

- For Thickness: Enter **.060 in**.

- Click **OK**.

19. Sketching the split profile: (to split the part into 2 halves)

- Select the <u>Right</u> plane from the FeatureManager tree.

- Click or select **Insert / Sketch**.

- Sketch a **3-Point-Arc** and add dimension as shown.

- Add a **Mid-point** relation between the end points of the arc and the outer-most edges.

Midpoint —

Right Plane

R80.000

— Midpoint

20. Removing the Upper Half:

- Click or select **Insert / Cut / Extrude**.

- Extrude Type: **Through All Both** (use Flip-side-to-Cut if needed).

- Direction 2:   **Through All** (default).

- Click **OK**.

The direction of the middle arrow indicates which half is being removed...

- The resulted cut, the lower half of the model.

21. Saving the lower half of the part:

- Select **File / Save As / Surface Loft Lower-Half / Save**.

---

22. Saving the lower half of the part:

- Select **File / Save As**, and enter: **Surface Loft Lower Half** for the file name.

- Enable the **Save As Copy and Continue*** check box and click **Save**.

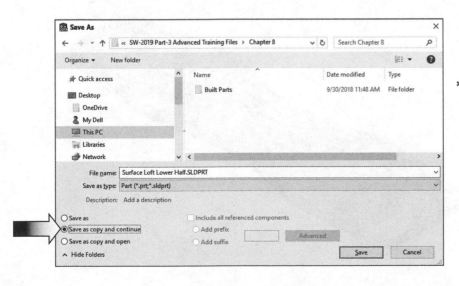

* This saves an exact copy of the same part but with a different name, so that we still have a full feature tree to create the second half of the part.

23. Modifying the copied file:

- Select **File / Open**.

- Select the document **Surface Loft Lower Half** and click **Open**.

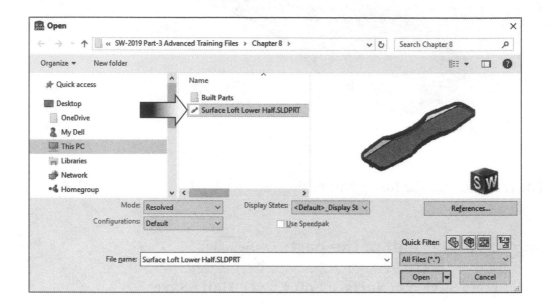

24. Changing the direction of the cut:

- Right click the **Cut-Extrude1** (the last feature on the tree) and select:

Edit Feature .

- Select **Flip Side To Cut** option ☑ Flip side to cut .

- Click **OK**.

The middle arrow pointing downwards indicates the lower half is being removed.

- The Lower Half of the part is removed, leaving the Upper-Half as the result of the Flip-Side cut.

25. Saving the upper half:

- Select **File / Save As**. Enter: **Surface Loft Upper Half** for the file name and click **Save.**

OPTIONAL:

- Insert the 2 halves into an assembly document and assemble them as shown.

Questions for Review

Using Surfaces

1. There are no limits on how many sections you can have in a loft feature.
   a. True
   b. False

2. Each loft section should be modeled onto a different plane.
   a. True
   b. False

3. The guide curves for use in a loft feature must be Coincident or Pierced to the sections.
   a. True
   b. False

4. Surfaces cannot be mirrored as solid features.
   a. True
   b. False

5. Only two surfaces can be used for knitting at a time.
   a. True
   b. False

6. Fillets cannot be used with surfaces, only in solid models.
   a. True
   b. False

7. Surfaces can be thickened after they are knitted together.
   a. True
   b. False

8. Mass properties options such as volume, surface area, etc., are available for all surfaces.
   a. True
   b. False

9. Surfaces can be knitted into a closed volume and then thickened into a solid.
   a. True
   b. False

	9. TRUE
8. FALSE	7. TRUE
6. FALSE	5. FALSE
4. FALSE	3. TRUE
2. TRUE	1. TRUE

<u>Exercise</u>: Loft & Delete Face

1. Create 2 new Planes, offset from Top plane.

2. Sketch the 1st Profile on Plane2.

3. Sketch the 2nd Profile on Plane1.

4. Sketch the 3rd Profile on Top plane
& add the connector points.

5. **Solid-Loft** the Profiles.

6. Add the Raised features (any
size), use Delete Face command and
remove 3 Faces (top, bottom & back).

CHAPTER 9

Offset Surface & Ruled Surface

Offset Surface & Ruled Surface

The Offset Surface command creates a new surface from a *single face* or a *set of faces*, with a distance of zero or greater. The offset surface can be created inward or outward.

Offset Surface

Offset from a single face   *Offset from a set of faces*

The Ruled Surface command creates a new surface from a single edge or a set of edges. The ruled surface can either be perpendicular or tapered from the selected edges.

Ruled Surface

Ruled surface from a single edge   *Ruled surface from a set of edges*

The Offset Surface and the Ruled Surface are used to create reference surfaces that help define the solid features in a part. In most cases, these surfaces should be knitted together before the next operation of extruded cuts, fillets, etc., can be performed.

Offset Surface & Ruled Surface

Dimensioning Standards: **ANSI**
Units: **INCHES** – 3 Decimals

Tools Needed:

 Lofted
Boss/Base

 Split Line

 Offset Surface

 Ruled Surface

 Knit Surface

 Shell

Using Offset & Ruled Surface options

1. Opening the existing file:

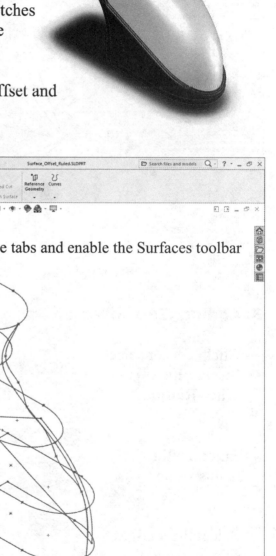

- Browse to the Training Files folder and open the part document named:
Surface_Offset_Ruled.sldprt

- This part document contains several sketches to be used as the Loft Profiles, and 2 other sketches used as the Guide Curves to help control the transition between each profile.

- This case study focuses on the use of the Offset and Ruled Surface commands.

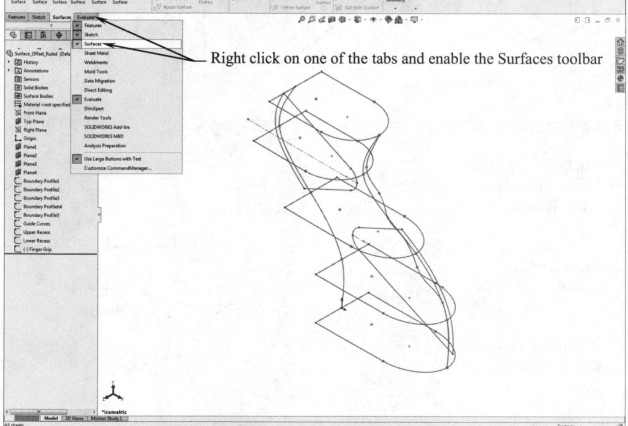

Right click on one of the tabs and enable the Surfaces toolbar

2. Creating the Base Loft:

- Click or select **Insert / Boss-Base / Loft**.

- Select the **5 Loft Profiles** and the **2 Guide Curves** as noted. (The SelectionManager appears when disjointed or overlapped entities are found in the sketch.)

2 Guide Curves. (Click OK in the Selection Manager after each sketch)

5 Loft Profiles

Guide Curve

Profile(Sketch1)

- Click **OK**.

3. Adding .250" fillets:

- Click or select **Insert / Features / Fillet-Round**.

- Enter **.250in**. for radius value.

- Select the **4 edges** shown to add the fillets.

- Click **OK**.

Radius: 0.25in

Fillet 4 edges

- The Split Line command projects an entity
to the surface(s) and divides the surface into
multiple faces.

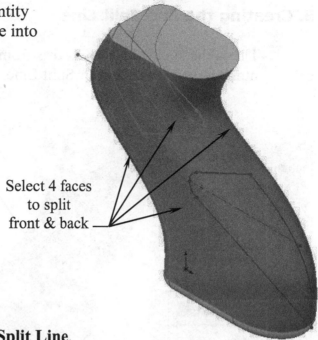

Select 4 faces
to split
front & back

4. Creating the 1st Split Line:

- Click or select **Insert / Curves / Split Line**.

- For Type-of-Split, select
the **Projection** option
(Default).

- For Sketch-To-Project,
select the sketch **Upper
Recess** from the Feature-
Manager tree.

- For Faces-To-Split,
select the **4 faces** as indicated.

- Click **OK**.

Select 4
Faces
Front &
back

- The handle body now has a new
set of faces, which will be used to
create the two recessed areas in the
next few steps.

5. Creating the 2nd Split Line:

- Using the sketch **Lower Recess** from the FeatureManager tree, repeat step number 3 to create the 2nd Split Line.

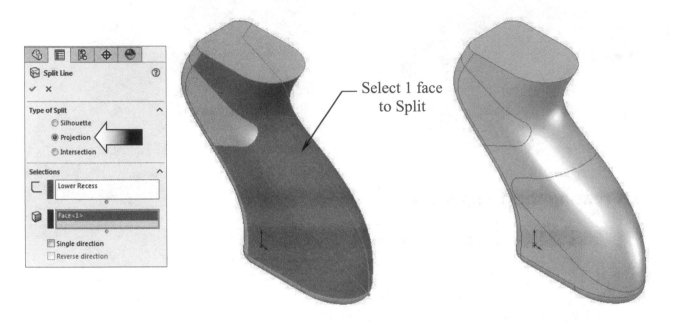

Select 1 face to Split

6. Creating the 1st Surface-Offset:

- Click or select **Insert / Surface / Offset**.

- Select the **5 Split-Faces** to offset.

- Enter **.050in**. for Offset Distance and click the **Reverse** button (Inside).

Offset 5 faces to the INSIDE

- Click **OK**.

7. Hiding the Solid Body:

- Right click on the **upper surface** of the solid body and select **Hide** 👁 .

8. Creating the 1ˢᵗ Ruled Surface:

- Click 🗗 on the Surfaces tool tab or select: **Insert / Surface / Ruled Surface**.

- Select the **Normal-To-Surface** option.

- Enter **.250in**. for Offset Distance.

- Right click on one of the **outer edges** and select: **Select-Open-Loop**.

- Click **OK**.

- The resulted Ruled Surfaces.

> ### Ruled Surfaces
>
> The Ruled Surfaces creates a set of surfaces that either perpendicular or taper from the selected edges.
> These surfaces can also be used as the Interlock Surfaces in molded parts

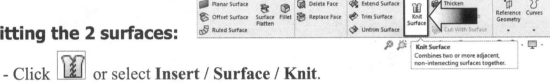

9. Knitting the 2 surfaces:

- Click or select **Insert / Surface / Knit**.

Knit Surface
Combines two or more adjacent, non-intersecting surfaces together.

- Select the **Surface-Offset** and the **Ruled-Surface** to knit.

- Click off the **Gap Control** checkbox (arrow).

- Click **OK**.

Knit two surfaces

10. Adding .020" Fillets:

- Click or select: **Insert / Features / Fillet-Round**.

- Enter **.020in**. for Radius value.

- Select all **Inner Edges** to fillet.
  (Right click / Select Tangency).

- Click **OK** .

Select all
Inner edges

- The resulting fillets.

11. Showing the Solid Body:

- From the FeatureManager tree, expand the Solid Bodies folder, right click on the **Split-Line2** body, and select: **Show Solid Body**.

12. Creating the Surface Cut:

- Click or select **Insert / Cut / With Surface**.

- Select the **Surface-Knit1** either from the graphics area or from the FeatureManager tree.

Direction arrow

- The Direction Arrow indicates the side that will be removed by the cut. Click Reverse (arrow) if needed.

- Click **OK**.

13. Hiding the Knit Surface:

- Right click on the Knit Surface and select **Hide** 👁.

Right click
and Hide

14. Creating the 2ⁿᵈ Ruled Surface:

- Repeat from step number 8 to create the 2ⁿᵈ Ruled Surface.

- Create the Surface Cut as indicated in step 12.

15. Adding more Fillets:

- Click or select:
Insert / Features / Fillet-Round.

- Enter **.020in**. for radius size.

- Select the <u>outer edges</u> of both upper and lower cuts to add the fillet.

- Click **OK**.

Select all outer edges

16. Shelling the part:

- Select the uppermost face as noted.

- Click or select **Insert / Features / Shell**.

- Enter **.020in**. for thickness.

- Click **OK**.

Select face to Shell

17. Saving your work:

- Click **File / Save As**.

- Enter: **Surface_Offset_Ruled** for file name.

- Click **Save**.

Exercise: Advanced Surfaces

1. Opening an Existing file:

- Browse to the Training Files folder.

- Open the document named:
  Advanced Surfaces Exercise.

2. Creating the Loft body:

- Create a **Solid Loft** from the **4 profiles** as indicated.

- Use the **2 bottom Guide Curves** to control the sides.

- Use the top **Guide Curve** to control the upper curvatures.

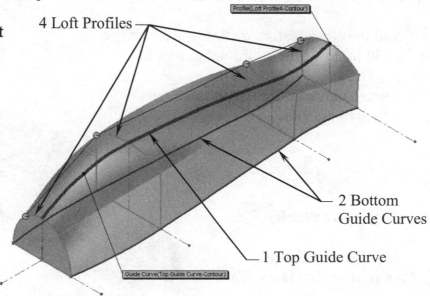

4 Loft Profiles

Profile(Loft Profile4-Contour)

2 Bottom Guide Curves

1 Top Guide Curve

Guide Curve(Top Guide Curve-Contour)

3. Adding Fillets:

- Add a **.500 in**. fillet to the upper edges, and a **.250 in**. fillet to the edges on the end, as indicated.

* By adding the fillets in the sketches the tangent lines can be eliminated by enabling the Merge-Tangent-Faces in the loft options.

R.500*

R.250

4. Creating the Split Lines & Lofted-Cuts:

- Use the 2 sketches named: **Circular-Split** and **Side-Split** to create 2 split surfaces.

- Create 2 lofted-cuts at **.090** in. deep, using either the **Offset** or **Ruled** surface options.

5. Adding the Nose and Fillets:

- Add the Nose feature that measured between **1.250 in**. to **1.500 in**. from the front face.

- Remove all sharp edges of the cutouts with **.040 in**. fillets.

6. Saving your work:

- Save the exercise as:
  Advanced_Surfaces_Exe.

Exercise: Advanced Surfacing Techniques

Surfaces

- Surfaces are a type of geometry that can be used to create solid features. Surfaces can be created in a variety of different ways; from a sketch or multiple sketches, a surface can be made by extruding, revolving, sweeping and lofting.

- Surfaces are normally created individually and knitted together so that an enclosed volume or a solid feature can be generated afterwards.

- This exercise discusses some advanced techniques on surfacing such as Lofted Surface, Boundary Surface, Trimmed Surface, Offset Surface, Extrude From, and variable fillets.

1. Opening the existing document named:

Advanced Surfacing Techniques from the Training Files folder. Rollback below the sketch: 1st Loft Guide-Curves.

2. Creating the 1st Lofted Surface:

- Click [icon] or select:
Insert / Surface / Loft.

- Select the **2 Loft Profiles** and the **2 Guide Curves** as indicated.

- Click **OK**.

Loft Profile 2

Guide Curve 1

Profile(1st Loft-Profile1)

Guide Curve

Loft Profile 1

Guide Curve 2

Lofted-Surface creates a feature by making transitions between two or more profiles. A loft can either be a surface or solid and one or more Guide Curves can be used to guide the transitions between the profiles.

3. Creating the 2ⁿᵈ Lofted Surface:

- Click or select: **Insert / Surface / Loft**.

- Select the **3 Loft Profiles** and the **4 Guide Curves** as indicated.

- Click OK after each selection of profile and guide curves.

3 Loft Profiles

3 Loft Profiles

4 Guide Curves

4 Guide Curves

Note: Click ✓ after each selection

Select Open Loop
Select Closed Loop
Select Region
Select Group

- Click **OK**.

4. Creating the Boundary-Surfaces:

- Click or select **Insert / Surface / Boundary Surface**.

- For **Direction 1**, select the **2 edges** as shown (Blue tags).

- For **Direction 2**, select the **3 Arcs** in the Boundary Sketch as shown (Purple tags).

> **Boundary-Surface** creates a new surface from a set of 2D or 3D sketch entities.
> The Boundary Surface can be tangent or curvature-continuous in both directions (all sides of the surface). This option offers a higher quality result than the Loft.

Direction 1:
Select 2 edges
(Blue Tags)
May need to select twice.

Blue

Purple

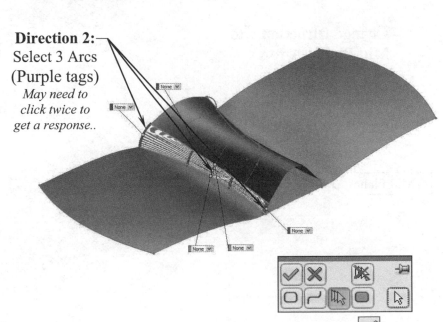

Direction 2:
Select 3 Arcs
(Purple tags)
May need to click twice to get a response..

Note: Click ✓ after each selection

- Click **OK**.

5. Repeating:

- Repeat step number 4 and create another Boundary Surface on the opposite side.

6. Creating an Extruded-Surface:

- Select the **Eyes Sketch** from the FeatureManager tree.

- Click [icon] or select: **Insert / Surface / Extrude**.

- Change the option **Extrude From** to **Surface/Face/Plane** (Arrow).

- Select the surface as indicated to extrude.

> **Extruded-Surface**
> creates a new surface from a 2D or 3D sketch, which protrudes normal to the sketch plane.

Extrude From

Select Surface

- Change **Direction 1** to **Mid Plane** (Arrow).

- Enter **.410in.** for extrude depth.

Mid Plane

- Click **OK**.

7. Creating a Trimmed Surface:

- Click [icon] or select **Insert / Surface / Trim**.

> **Trimmed-Surface** uses a plane, a surface, or a sketch as a trim tool to trim the intersecting surfaces.

- Select **Mutual** under Trim Type (Arrow).

- For **Trimming Surfaces**, select **all surfaces** of the model.

- For **Remove Selections**, select the **5 Faces** as shown (Arrow).

Pieces to Remove

Select Mutual

Trimming Surfaces

Remove Sel.

Pieces to Remove

8. Creating another Extruded Surface:

- Select the **Thunder Sketch** and click [icon] or select **Insert / Surface /Extrude**.

- Set the **Direction 1** to **Blind** and click **Reverse Direction**.

- Set **Extrude Depth** to: **1.250in**.

- Click **OK**.

9. Creating an Offset-Surface:

- Click or select **Insert / Surface / Offset**.

- Select the **2 surfaces** as shown (arrow).

- Enter **.100in.** for **Offset Distance**.

- Place the copy on the **bottom** of the original.

Note: Create 2 offset surfaces separately if the next trim failed.

- Click **OK**.

> **Offset-Surface** creates a copy of a surface in either direction, and is parallel to the selected surface(s). The offset distance can be zero or any other value.

Select 2 surfaces to offset...

Offset Surface .100in. below...

10. Creating a Mutual Trimmed Surface:

- Click or select **Insert / Surface / Trim***.

- Select **Mutual** under Trim Type (Arrow).

- For **Trimming Surfaces**, select **all surfaces** of the Thunder Sketch and the Offset Surfaces.

Trimming Surfaces

Trimming Surfaces

* Note: Using Standard-Trim to trim one feature each time would also work well.

- For **Remove Selection**, select the following:

* The **surfaces** of the **Thunder Sketch**; keep the inside faces as noted.

* The **2 Offset Surfaces**.

Keep the inside surfaces…

Remove Sel.

- Hide all trimmed surfaces when done.

- Click **OK**.

11. Creating an Extruded Surface :

- Select the **Corner Radius Trim** Sketch and click or select:
  Insert / Surface / Extrude.

- Set the **Direction 1** to **Blind** and click **Reverse Direction**.

- Set Depth to **2.000in**.

- Click **OK**.

(These new surfaces will be used to trim-off the 4 corners).

12. Creating a corner Trimmed Surface:

- Click or select **Insert / Surface / Trim**.

- Select **Mutual** under Trim Type (arrow).

- For **Trimming Surfaces**, select the following:

> * The **4 extruded faces** and the **4 corner pieces**.

> * The **2 left/right faces** of the model.

Trimming Surfaces

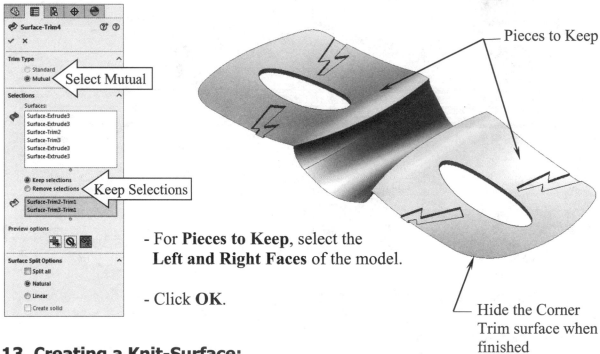

Pieces to Keep

Select Mutual

Keep Selections

- For **Pieces to Keep**, select the **Left and Right Faces** of the model.

- Click **OK**.

Hide the Corner Trim surface when finished

13. Creating a Knit-Surface:

- Click or select **Insert / Surface / Knit**.

- Select **all surfaces** either from the FeatureManager tree or from the graphics area.

- Click off the Gap Control.

- Click **OK**.

14. Adding a Variable Fillet:

- Click or select **Insert / Features / Fillet-Round**.

- Select the **2 edges** shown.

- Use the **Call-out tags** and enter the radius values as noted.

- Click **OK**.

15. Adding a Constant Fillet:

- Click or select **Insert / Features / Fillet-Round**.

- Select the **2 edges** as shown.

- Enter **.200in.** for Radius values.

- Click **OK**.

16. Optional: Adding texture

- Enable the **RealView Graphics** option.

- From the Task Pane expand the **Appearances** folder, and locate the **Rubber / Texture** folder.

- Drag & Drop the **Textured Rubber** onto the part, and select the Apply to Part option .

Textured Rubber

17. Saving your work:

- Click **File / Save As**.

- Enter **Advanced Surfacing Techniques** for the name of the file.

- Click **Save**.

CHAPTER 10

Advanced Surfaces

Advanced Surfaces
Computer Mouse

Surfaces are a type of geometry that can be used to
create solid features. Surface tools are available on the Surfaces toolbar.

You can use surfaces in the following ways:

* Select surface edges and vertices to use as a sweep guide curve and path.
* Create a solid or cut feature by thickening a surface.
* Extrude a solid or cut feature with the end condition Up to Surface or Offset from Surface.
* Create a solid feature by thickening surfaces that have been knit into a closed volume.
* Replace a face with a surface.

Surface body is a general term that describes connected zero-thickness geometries such as single surfaces, knit surfaces, trimmed and filleted surfaces, and so on. You can have multiple surface bodies in a single part.

Keep in mind that when modeling with surfaces, every feature should be easily edited. One way to achieve that is to try and create one surface at a time, and towards the end, knit all surfaces into one prior to converting it into a solid model.

This handout will show us one of the more convenient methods where multibody surfaces are used to create the computer mouse body. Each surface will be created individually and then trimmed and knitted into a single surface body.

After the surface knit, a wall thickness can be added to the surface body to convert it into a solid body so that other features can be added to it from that point.

Advanced Surfaces
Computer Mouse

Dimensioning Standards: **ANSI**
Units: **INCHES** – 3 Decimals

Tools Needed:

 Extruded Surface

 Trim Surface

 Mirror Surface

 Filled Surface

 Planar Surface

 Knit Surface

1. Starting a new part document:

- Click **File / New**.

- Select the **Part** template and click **OK**.

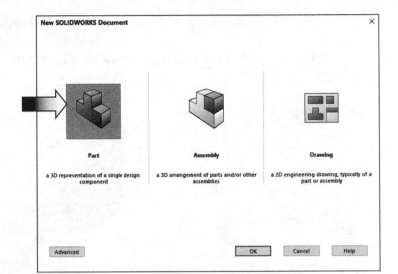

2. Sketching the 1st profile:

- Select the **Top** plane and open a new sketch.

- Sketch a **Centerline** and a **3-Point Arc** shown below.

- Add the dimensions and relations to fully define the sketch.

3. Extruding the 1st surface:

- Switch to the **Surfaces** toolbar and click **Extruded Surface** .

- Use the default **Blind** type and enter a Depth of **1.00in**.

- Click the **Draft On/Off** button and enter **3.00deg** for Draft Angle.

- Click **OK**.

4. Sketching the 2nd profile:

- Select the **Front** plane and open a new sketch.

- Sketch a **3-Point-Arc** and add the dimensions / relations as indicated.

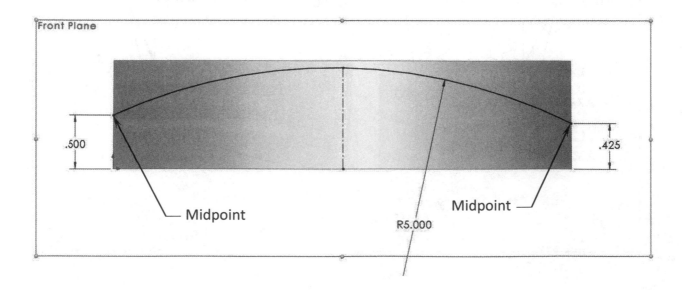

5. Extruding the 2nd surface:

- Switch to the **Surfaces** toolbar and click **Extruded Surface** .

- Use the default **Blind** type and enter a Depth of **1.50in**.

- Click **OK**.

6. Trimming the surfaces:

- Select the **Trim Surface** command .

- Use the default Standard Trim option. For Trim Tool, select the **Surface-Extrude2**.

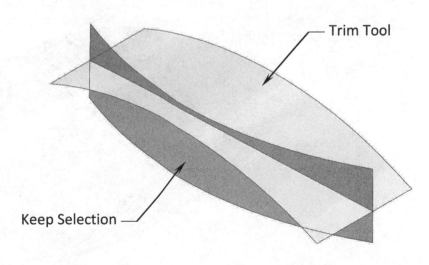

Trim Tool

Keep Selection

- For Keep Selections, select the <u>lower portion</u> of the **Surface-Extrude1** as indicated.

- Click **OK**.

7. Mirroring the surfaces:

- Switch to the **Features** tool tab.

- Click the **Mirror** command.

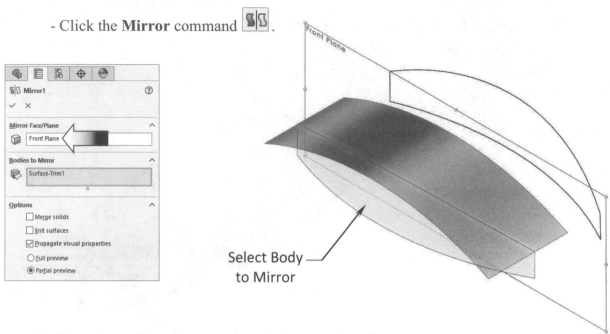

- For **Mirror Face/Plane**, select the **Front** plane from the FeatureManager tree.

- Expand the **Bodies to Mirror** section and select the lower portion of the **Surface-Trim1** as noted.

- Click **OK.**

- Your surface model should look like the image above.

8. Creating the 1st Filled Surface:

- Switch back to the **Surfaces** tool bar.

- Select the **Filled Surface** command .

Filled Surface

The Filled Surface feature constructs a surface patch with any number of sides, within a boundary defined by existing model edges, sketches, or curves, including composite curves.

- For Patch Boundary, select the top **4 edges** as noted.

- Expand the Constraint Curves section and select the **3D Sketch Curve** as indicated, to help control the curvature of the patch.

- Leave all other parameters at their default settings.

- Click **OK**.

- Compare the result of your Filled Surface to the image shown here.

9. Creating the 2nd Filled Surface:

- Rotate the right end of the model to a similar position shown below.

- Click the **Filled-Surface** command again.

- For Patch-Boundary, select the **4 edges** on the right end as indicated.

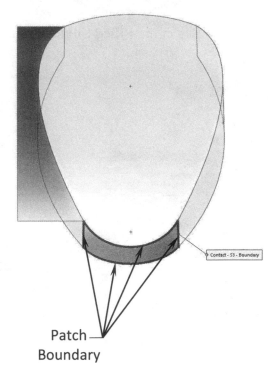

Patch Boundary

- Click **OK**.

10. Creating the 3rd Filled Surface:

- Select the **Filled Surface** command once again.

- Rotate the left end of the model approximately as shown below.

- For Patch Boundary, select the **4 edges** as noted in the image.

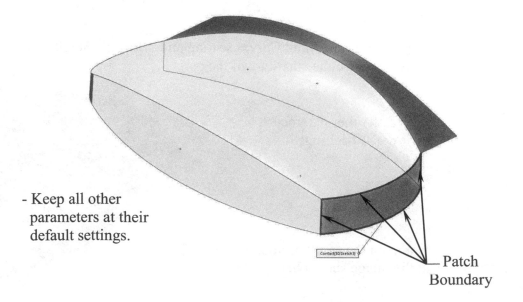

- Keep all other parameters at their default settings.

Patch Boundary

- Click **OK**.

- For clarity, **hide** the surface body indicated below.

Hide this surface

11. Creating a Planar Surface:

- Rotate the surface model so that the bottom opening is visible.

- Select the **Planar Surface** command .

Bounding Entities

- For Bounding Entities, select the **4 edges** on the bottom of the surface model.

- The preview of a planar surface showing the bottom opening is being closed off.

- Click **OK**.

12. Creating a Knit Surface:

- Click the **Knit Surface** command 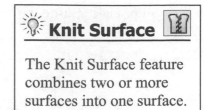.

- Select all **6 surfaces** as noted.

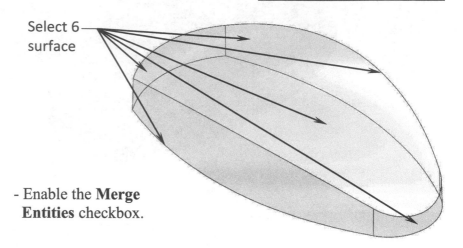

Select 6 surface

- Enable the **Merge Entities** checkbox.

- Click **OK**.

> **Knit Surface**
>
> The Knit Surface feature combines two or more surfaces into one surface.

13. Creating the 1.00" fillet:

- Select the **Fillet** command on the Surfaces tool bar.

- Use the default **Constant Size** fillet.

- Select the **2 edges** shown.

- Enter **1.00in** for radius.

Select 2 edges

- Click **OK**.

14. Creating the .040" fillet:

- Select the **Fillet** command again.

- For radius size, enter **.040in**

- For Items to Fillet, select the **bottom edge** of the model.

- The Tangent-Propagation should be enabled.

- Click **OK**.

15. Creating the .400" fillet:

- Select the **Fillet** command once again.

- For Items to Fillet, select the **top edge** of the surface model.

- For radius size, enter **.400in**

- Enable the Tangent-Propagation checkbox to allow the fillet to propagate all around the bottom of the surface model.

- Click **OK**.

16. Saving your work:

- Click **File / Save As**.

- For the file name, enter:
  Computer Mouse.sldprt

- Click **Save**.

Shaded with Edges

Shaded with Ambient Occlusion (Back Iso)

Shaded with Ambient Occlusion (Front Iso)

Using Filled Surfaces

Using Filled Surface
Patch with Curvature Controls

Use Filled Surface and Planar Surface commands to fill or patch a surface boundary with any number of sides.

The boundary can be a set of existing model edges, sketches, or curves, including composite curves. The boundary should be closed for the patch to work properly.

There are several options to help you control the curvatures when patching a surface boundary such as Contact, Tangent, and Curvature. These options are explained later in the lesson.

Other than the Curvature Control options the Apply-to-All-Edges check box enables you to apply the <u>same</u> curvature control to all edges. If you select the function after applying both **Contact** and **Tangent** to different edges, it applies the current selection to <u>all</u> edges.

If your surface model has two or four-sided surfaces, try using the **Optimize surface** option. The Optimize surface option applies a simplified surface patch that is similar to a lofted surface. Potential advantages of the optimized surface patch include faster build times, and increased stability when used in conjunction with other features in the model.

This lesson will teach us the use of the Filled Surface and Planar Surface commands.

Using Filled Surfaces
Patch with Curvature Controls

Dimensioning Standards: **ANSI**

Units: **INCHES** – 3 Decimals

Tools Needed:

 Planar Surface   Filled Surface   Knit Surface

1. Opening a part document:

- Click **File / Open**.

- Browse to the Training Files folder,
  open the part document named:
  Filled Surfaces.sldprt

- This part document was previously
  saved as a different file format. There is
  no feature history available on the FeatureManager tree.

- There are three openings in the part that we will have to fill using different
  options within the Filled-Surface command.

- If the Feature Recognition dialog pops up, click **NO** to close it.

2. Enabling the Surfaces toolbar:

- Right click the Features tab and select the **Surfaces** tool tab.

Right click here and enable the Surfaces toolbar

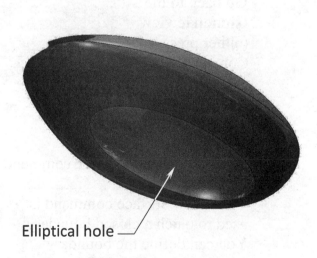

Elliptical hole

- Rotate the model and locate the elliptical-hole at the bottom of the model.

- The opening should form a flat surface and we can use the Planar-Surface
  command to patch it.

3. Creating a Planar surface :

- Click the **Surfaces** tool tab (arrow).

- Select the **Planar-Surface** command.

- Select the edge of the elliptical hole.

- The preview of a new surface appears.

- Click **OK**.

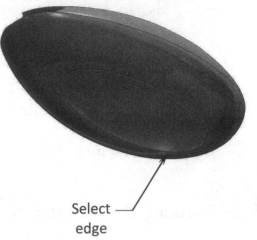

- A planar surface can be created from a sketch, a set of closed edges, or a pair of planar entities such as curves or edges.

Select edge

4. Creating a Surface Fill with Tangent Control :

- Go back to the **Isometric view** (either press Control + 7 or press the Spacebar and select the Isometric view).

- Click the **Filled Surface** command.

- The Filled Surface command is used to patch a closed boundary. You can define the boundary by selecting a set of 2D or 3D sketch entities, model edges or composite curves.

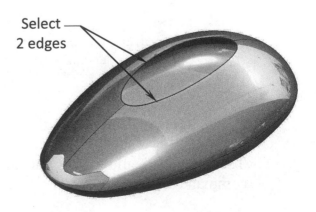

Select 2 edges

- Select the two edges of the elliptical hole as indicated.

- A preview mesh is displayed to help you
  visualize the curvature
  of the new surface.

- Change the
  Curvature Control*
  to **Tangent**.

- Enable the
  checkboxes as
  shown.

- Click **OK**.

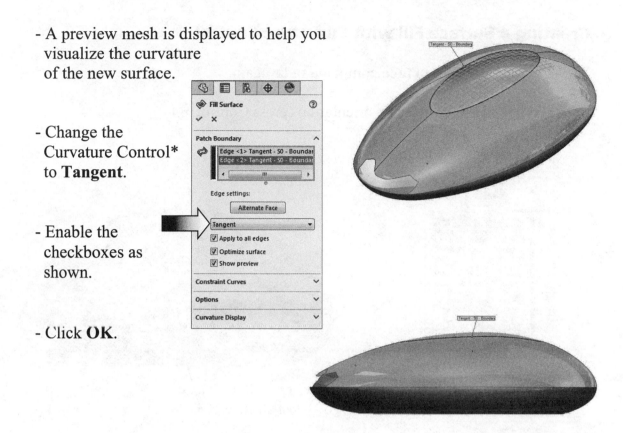

* The **Curvature Control** defines the type of control you want to exert on the
  patch you create. The types of **Curvature Control** include:

  - **Contact**: Creates a surface within the selected boundary.

  - **Tangent**: Creates a surface within the selected boundary, but maintains
    the tangency of the patch edges.

  - **Curvature**: Creates a surface that matches the curvature of the selected
    surface across the boundary edge with the adjacent surface.

5. Creating a Surface Fill with Curvature Control 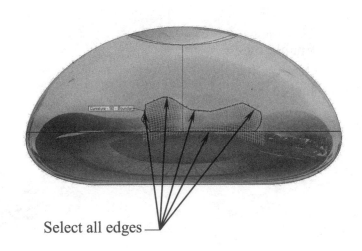 :

- Click the **Filled Surface** command once again.

- Change to the Front view orientation (press Control + 1).

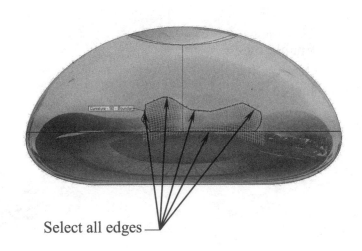

Select all edges

- Select <u>all edges</u> of the opening in the front as noted.

- The preview mesh appears indicating a closed boundary is found. (Enable the Preview Mesh checkbox if the preview is not visible.)

- Under the Curvature Control change the Contact option to **Curvature** (arrow).

- Enable the other checkbox as shown in the dialog box.

- Click **OK**.

6. Knitting all surfaces :

- At this point all openings have been filled. We can now combine all surfaces into one by using the Surface-Knit option.

- Click the **Knit Surface** on the Surfaces tool tab.

Select 4 surfaces

- Select all for surfaces in the graphics area as indicated.

- Click the **Try To Form Solid** checkbox (arrow) to convert the part to a solid model.

- Click **OK**.

- To verify the interior of the part, create a section view using the **Right** plane as the cutting plane. Click Cancel when done.

7. Saving your work:

- Save your work as **Using Filled Surface**.

Patch Types:

A closer look at the Edge Settings: Use the Edge Settings options to define the type of control you want to exert on the patch you create.

Contact Patch:

Creates a surface within the selected boundary.

Tangent Patch:

Creates a surface within the selected boundary, but maintains the tangency of the patch edges.

Curvature Patch:

Creates a surface that matches the curvature of the selected surface across the boundary edge with the adjacent surface.

Optional:

A closer look at the Zebra Stripes:

Use zebra stripes to visually determine which type of boundary to use between surfaces.

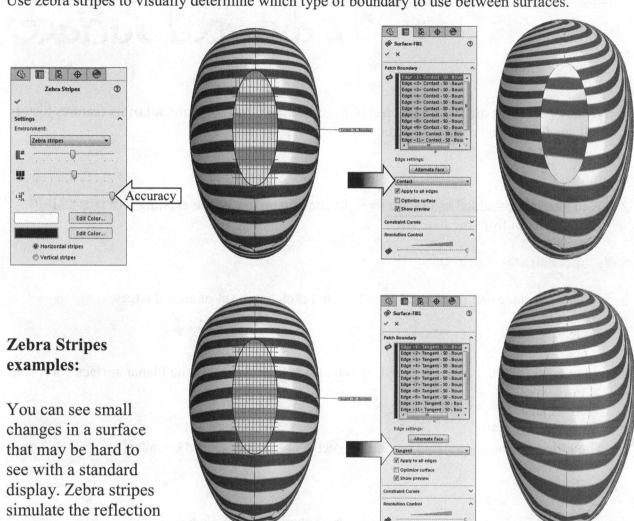

Zebra Stripes examples:

You can see small changes in a surface that may be hard to see with a standard display. Zebra stripes simulate the reflection of long strips of light on a very shiny surface.

With zebra stripes, you can easily see wrinkles or defects in a surface, and you can verify that two adjacent faces are in contact, are tangent, or have continuous curvature.

Questions for Review

Advanced Surfaces

1. When opening a part document created from another CAD software, all of its features history will appear on the FeatureManager tree.
    a. True
    b. False

2. Tool tabs can be added as needed by right clicking on one of the existing tool tabs and selecting them from the list.
    a. True
    b. False

3. A planar surface can be created from a closed sketch or a set of closed edges.
    a. True
    b. False

4. An open sketch or a set of open edges can also be patched using the Planar surface command.
    a. True
    b. False

5. The Filled Surface command is used to patch a non-planar closed boundary.
    a. True
    b. False

6. The Filled Surface command will fail if the boundary is not closed.
    a. True
    b. False

7. The Preview Mesh can be toggled on/off during the creation of the filled surface.
    a. True
    b. False

8. The Knit Surface command can only Knit the surfaces; it cannot form a solid from the surfaces.
    a. True
    b. False

7. TRUE
5. TRUE
3. TRUE
1. FALSE

8. FALSE
6. TRUE
4. FALSE
2. TRUE

Boundary and Freeform Surfaces

Boundary Surface: Allows the user to create surfaces that can be tangent or curvature continuous in both directions (all sides of the surface). In most cases, this delivers a higher quality result than the loft tool.

Freeform feature: Modifies a face of a surface or a solid body. You can modify only one face at a time and the face can have any number of sides. Control curves and control points can be added to allow pushing and pulling the control points to modify the face. The triad is used to constrain the push or pull direction.

1. Opening a part file:

- Browse to the Training Files folder and open the part file named: **Freeform Surface**.

2. Creating the 1st Boundary Surface:

- Right click either the Features or the Evaluate tab and enable the **Surfaces** option.

- Click the **Boundary Surface** button from the Surfaces toolbar or select: **Insert / Surface / Boundary Surface**.

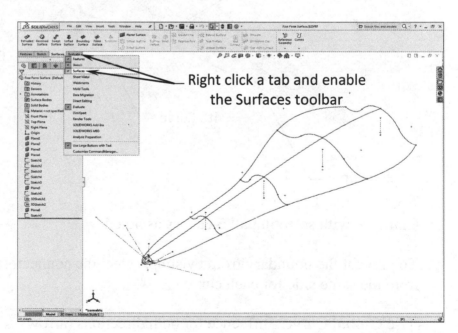

Right click a tab and enable the Surfaces toolbar

- Zoom in on the left end of the image.

- For Direction 1: select the **1st Curve** approximately as indicated.

  - Curves are used to determine the boundary feature in Direction 1 and Direction 2.

Click here to select the 1st Curve...

- For each curve, select the point from which you want the path of the boundary feature to travel.

5th Curve

3rd Curve

Connector

4th Curve

2nd Curve

- Continue with selecting all **5 curves** as noted.

- To prevent the boundary from twisting, select the connectors from the same side for each curve.

- Use **Global** Curves Influence for both directions (arrow) to extend the curve influence to the entire boundary feature.

- Continue on next page....

- For Direction 2: select the **3 Curves** as indicated below.

- Since both Curve6 and Curve7 were created in the same sketch, the **Selection–Manager** pops up when one of them is selected, asking to confirm your selection. Click the OK button after selecting Curve6 and Curve7.

6th Curve

Click OK after the
6th and 7th Curves

7th Curve

8th Curve

- Enable the **Merge Tangent Faces** option.

- Enable the Mesh Preview option and change the Mesh-Density to **12**.

- Click **OK**.

3. Creating the 2nd Boundary Surface:

- Zoom in on the tip section of the surface.

- There are 5 other Curves on the Feature-Manager tree; 2 of them will be used as Direction 1 and the other 3 used as Direction2.

- Click the **Boundary Surface** button from the Surfaces toolbar or select: **Insert / Surface / Boundary Surface**.

- For Direction 1, select the **Curve1** and **Curve2** as noted.

Curve 1

Curve 2

- For Direction 2, select the **Curve3** as indicated.

- Click OK in the **Selection-Manager** dialog box after selecting the Curve3.

- Continue on next page...

Curve 3

Click OK after the 3rd and 5th Curves

- For Direction2, continue with selecting the **Curve4** and **Curve5** as noted below.

- Enable the **Merge Tangent Faces** checkbox to force the boundary feature to be tangent to the corresponding segments.

Curve 5

- Enable the Mesh-Preview and change the Mesh-Density to **12**.

Curve 4

- Click **OK**.

- The resulting Boundary Surface.

4. Creating the Freeform feature:

- From the **Insert** menus, select **Features / Freeform**.

- The **Freeform** Property-Manager appears when you create a freeform feature. Only one face can be modified at a time, but the face can have any number of sides.

- A set of Control Curves and Control Points are used to modify the face.

- **Select the face** as indicated to modify.

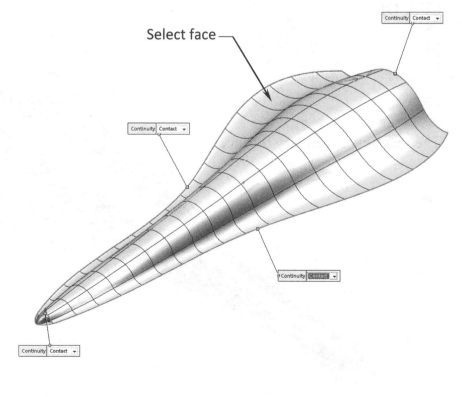

Select face

- In the **Control Curves** section click the **Add Curves** button.

- Position the cursor approximately at the center of the face and click to **add a curve**.

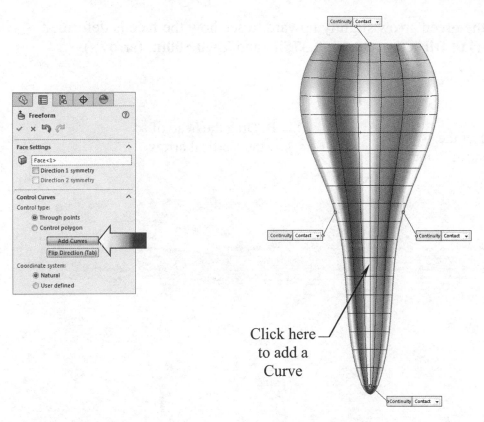

Click here
to add a
Curve

- In the **Control Points** section click the **Add Points** button.

- Show the sketch2, sketch3, and sketch4 to help with placing the points more easily.

- Change to the Right view orientation (Control+4) and **Add 3 points** along the curve approximately as indicated.

Add 3rd point

Add 2nd point

Add 1st point

- Click the Add Point button again to turn it off, after adding the 3rd point.

- Select the 1st Control Point to see the Triad. The arrows on the triad are used to push or pull the surface one direction at a time. Drag the center sphere in the middle of the 3 arrows to push or pull in all directions at the same time.

- Drag the head of the green arrow slightly upward to see how the face is deformed, then change the **Triad Directions** to **X= 0.375in.** and **Y= 0.500in.** (arrows).

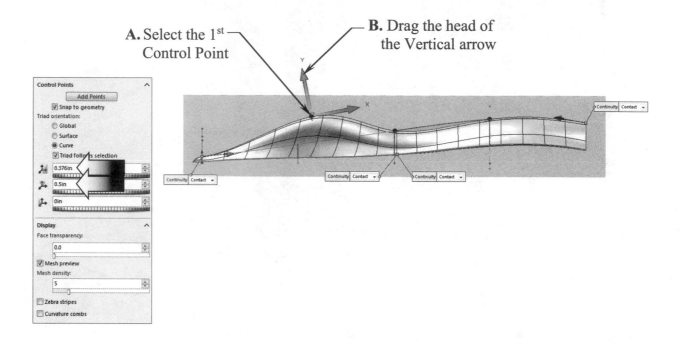

A. Select the 1st Control Point

B. Drag the head of the Vertical arrow

- Click the 2nd Control Point and set the **Triad Directions** to:
  X= -0.375" and **Y= 0.050"**.

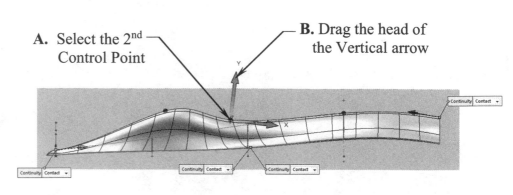

A. Select the 2nd Control Point

B. Drag the head of the Vertical arrow

- Click the 3rd Control Point; set the **Triad Directions** to **X= -0.700"** and **Y= 0.015"**.

A. Select the 2nd Control Point

B. Drag the head of the Vertical arrow

- Click the **Curvature combs** to see a visual enhancement of the surface' curvatures.

- Set the **Scale** to adjust the size of the curvature combs and set the **Density** to adjust the number of curvature combs.

- Click **OK**.

- The resulting Freeform Surface.

5. Saving your work:

- Click **File / Save As**.

- Enter **Boundary & Freeform** for the name of the file.

- Click **Save**.

- Overwrite the old file with the new, if prompted.

CHAPTER 11

Surfaces vs. Solid Modeling

Surfaces vs. Solid Modeling

Surfaces are a type of geometry that can be used to create solid features. Surface tools are available on the Surfaces toolbar.

You can <u>create surfaces</u> using these methods:

- Insert a planar surface from a sketch or from a set of closed edges that lie on a plane
- Extrude, revolve, sweep, or loft, from sketches
- Offset from existing faces or surfaces
- Import a file
- Create mid-surfaces
- Radiate surfaces

You can <u>modify surfaces</u> in the following ways:

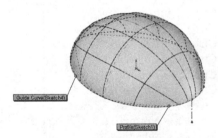

- Extend
- Trim existing surfaces
- Un-trim surfaces
- Fillet surfaces
- Repair surfaces using Filled Surface
- Move/Copy surfaces
- Delete and patch a face
- Knit surfaces

You can <u>use surfaces</u> in the following ways:

- Select surface edges and vertices to use as a sweep guide curve and path
- Create a solid or cut feature by thickening a surface
- Extrude a solid or cut feature with the end condition Up to Surface or Offset from Surface
- Create a solid feature by thickening surfaces that have been knit into a closed volume
- Replace a face with a surface

Surfaces vs. Solid Modeling
Safety Helmet

View Orientation Hot Keys:

Ctrl + 1 = Front View
Ctrl + 2 = Back View
Ctrl + 3 = Left View
Ctrl + 4 = Right View
Ctrl + 5 = Top View
Ctrl + 6 = Bottom View
Ctrl + 7 = Isometric View
Ctrl + 8 = Normal To
      Selection

Dimensioning Standards: **ANSI**

Units: **INCHES** – 3 Decimals

Tools Needed:

Insert Sketch	Plane	Lofted Surface
Swept Surface	Planar Surface	Knit Surface
Revolve Cut	Sweep Cut	Surface Thicken

1. Opening the Existing file:

- From the Training Files folder open the part document named: **Helmet**.

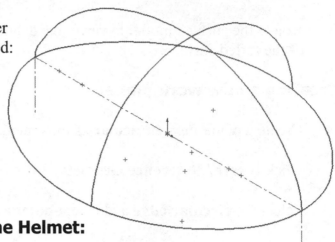

- This part file has 3 sketch profiles and 3 guide curves previously created.

2. Constructing the Body of the Helmet:

- Click **Lofted Surface** ![icon] OR – select **Insert / Surface / Loft**.

- Select the **3 Sketch Profiles**.

- Select the **3 Guide Curves** as noted.

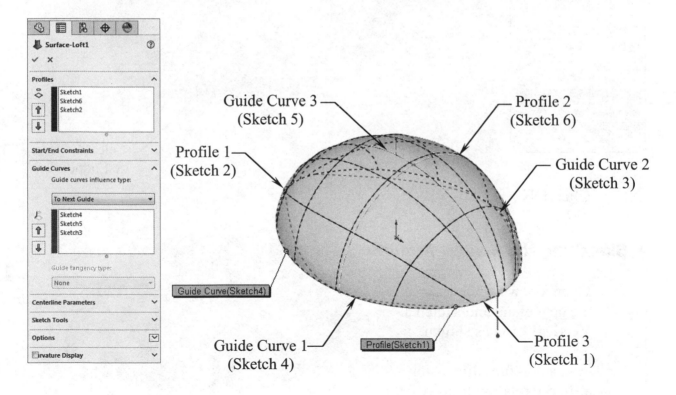

Guide Curve 3 (Sketch 5)

Profile 2 (Sketch 6)

Profile 1 (Sketch 2)

Guide Curve 2 (Sketch 3)

Guide Curve 1 (Sketch 4)

Profile 3 (Sketch 1)

Guide Curve(Sketch4)

Profile(Sketch1)

- Click **OK**.

- The resulted Surface-Loft.

- Rotate the surface model to verify the details of the lofted surface.

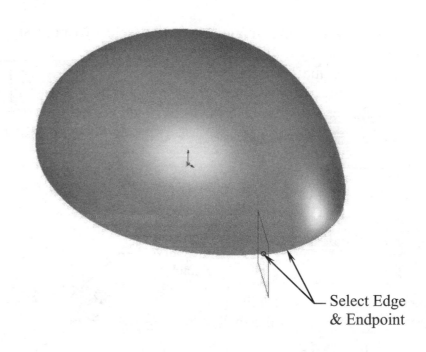

3. Creating a new work plane:

- Create a plane **Perpendicular** as illustrated.

- Click **Insert / Reference Geometry / Plane** .

- Select the **circular edge** and its **endpoint** as noted.

Select Edge & Endpoint

- Click **OK**.

4. Sketching the Sweep-Profile:

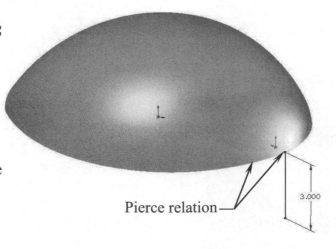

- Open a new sketch on the new plane and sketch a **Vertical Line** as shown.

- Add a **3.000in.** dimension and a **Pierce** relation to fully define the sketch.

- **Exit** the Sketch .

Pierce relation

3.000

5. Creating the Sweep-Path (Composite Curve):

- Click **Composite** on the Curve toolbar – OR –

select **Insert / Curve /Composite** 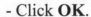.

- Select **all edges** (as noted).

- Click **OK**.

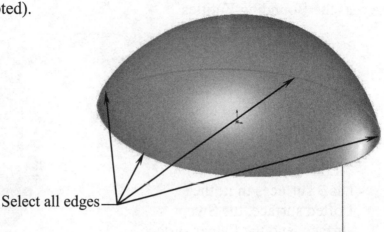

Select all edges

6. Creating a Swept-Surface

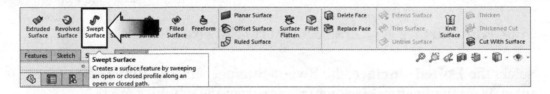

- Click **Swept Surface** – OR – **Insert / Surface / Sweep**.

- Select the **Vertical Line** as the Sketch Profile.

- Select the **Composite-Curve** as the Sweep Path.

- Select **Keep Normal Constant** under Options. (This option makes the section remain parallel to the beginning section at all times).

- Click **OK**.

7. Adding a Planar Surface:

- Click **Planar Surface** or select:
Insert / Surface / Planar.

- Select **all edges** on the bottom
as the Bounding Entities.

- Click **OK**.

- The new planar surface
is created and it covers
the bottom of the part.

- At this point, the model
has 3 surfaces in it: the
Lofted surface, the Swept
surface, and the Planar surface.

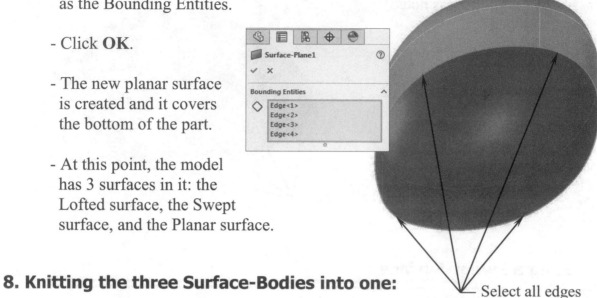

Select all edges

8. Knitting the three Surface-Bodies into one:

- Click **Knit Surface** or select **Insert / Surface / Knit**.

- Select the **Lofted-Surface**, the **Swept-Surface**, and the **Planar-Surface**
either from the FeatureManager tree (or from the graphics area).

- <u>Clear</u> the **Gap Control** checkbox.

- Enable the **Create Solid** option to convert the surface into a solid body.

- Click **OK**.

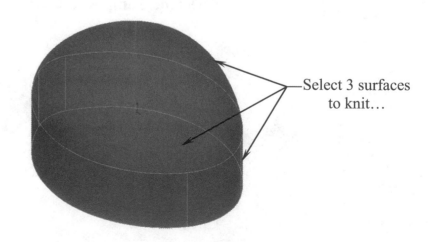

Select 3 surfaces
to knit…

9. Creating a section view:

- Select the <u>Front</u> plane from the FeatureManager and click **Section View** (arrow) to verify the solid material. Click off the section view command when finished viewing.

10. Adding an Extruded Cut feature:

- Select the **Sketch8** from the FeatureManager Tree.

- This sketch will be used to remove the lower portion of the Helmet.

- Click **Extruded-Cut** 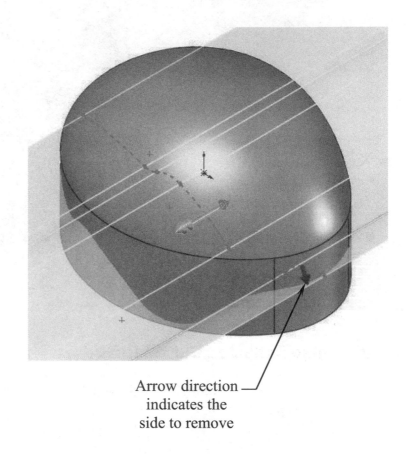 or select: **Insert / Cut / Extrude**.

- Select **Through-All Both** conditions from the drop-down list.

- Enable **Flip-Side-To-Cut** if needed to remove the lower portion of the part.

- Click **OK**.

Arrow direction ⎯
indicates the
side to remove

- <u>Hide</u> the Sketch8.

- Compare your model
with this illustration
before moving to the
next step.

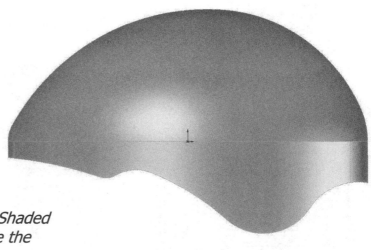

*<u>NOTE:</u> Change from Shaded to Shaded
With-Edges to be able to see the
edges more clearly.*

- The resulted cut.

11. Adding a Revolve-Cut feature:

- Select the **Sketch10** from the FeatureManager Tree.

- This sketch will be used to shape the <u>inside</u> of the Helmet.

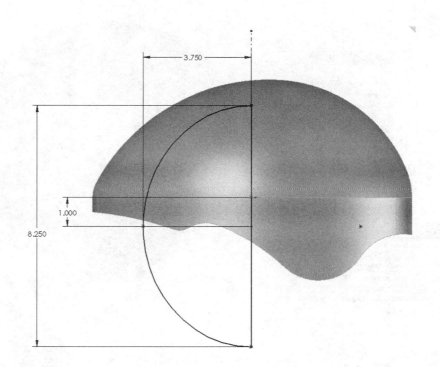

- Click **Revolve-Cut** or select **Insert / Cut / Revolve**.

- Select the **vertical centerline** as Revolve-Direction.

- Revolve Angle: **360.00deg**.

- Click **OK**.

- <u>Hide</u> the Sketch10.

12. Adding the side cut features:

- Select the **Sketch9** from the FeatureManager Tree.

- This sketch will be used to cut from the <u>inside</u> with a **1.00deg**. draft angle.

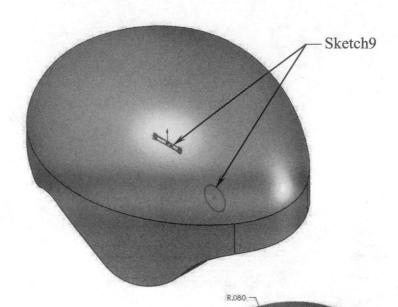

- Click **Extruded-Cut** or select: **Insert / Cut / Extrude**.

- Select **Through-All** for end condition.

- Enable Draft Angle and enter: **1.00deg**.

- Enable **Draft Outward** option.

- Click **OK**.

- <u>Hide</u> the Sketch9.

13. Creating the cut-out slot:

- Select the **Sketch11** from the FeatureManager Tree.

- This sketch will be used to cut a slot from the <u>outside</u> with a **10.00deg**. draft angle.

- Click **Extruded-Cut** or select: **Insert / Cut / Extrude**.

- For Direction 1, select **Up-To- Next** end condition.

- Enable Draft Angle and enter: **10.00deg**. (inward)

- For Direction 2, select **Through-All** end condition, no draft. (The 2nd direction is needed to remove the material on the mirrored side.)

- Click **OK**.

- <u>Hide</u> the Sketch11.

14. Mirroring the slot:

- Click **Mirror** – OR – select **Insert / Pattern-Mirror / Mirror** .

- For Mirror Plane, select the **Right** plane.

- For Features to Mirror, select the **Cut Extrude3**.

- Click **OK**.

15. Mirroring the Cut Features:

- Click **Mirror** or select: **Insert / Pattern-Mirror / Mirror**.

- For Mirror Plane, select the **Front** plane.

- For Features to Mirror, select **both Slots** and the **Side-Holes** as Features to Mirror.

- Click **OK**.

- Compare your model with the image below.

16. Creating the Sweep-Cut:

- Click **Insert / Cut / Sweep** .

- Select the sketch **Sweep-Profile** and the sketch **Sweep Path** from the FeatureManager tree.

- The preview graphics show the proper transition of the sweep feature.

Sweep Path

Sweep Profile

- Click **OK**.

17. Adding the .500" fillets:

- Click **Fillet** or select **Insert / Features / Fillet-Round**.

- Enter **.500in** for radius size.

- Select the **12 edges** of the Swept feature.

- Click **OK**.

Right click and Select Tangency

Select the 4 edges on 2 sides

18. Adding the .250" fillets:

- Click **Fillet** or select: **Insert / Features / Fillet-Round**.

- Enter **.250"** for radius size.

- Select **all edges** of the part (Control+A).

- Click **OK**.

19. Saving your work:

- Click **File / Save As / Helmet / Save**. (Save on the Desktop.)

Front Isometric

Back Isometric

Exercise: Advanced Loft – Turbine Blades

1. Open the existing document:

Turbine.sldprt from the Training Files folder.

- This part file has 6 sketch profiles and 4 guide curves previously created.

2. Create the 1st loft:
- Select the 3 Tall Blade sketches for Profiles.
- Select the 2 Tall Guide Curve sketches for Guide Curves.

3. Create the 2nd loft:
- Select the 3 Short Blade sketches for Profiles.
- Select the 2 Short Guide Curve sketches for Guide Curves.

The resulted Lofts.

4. Circular pattern the Blades:
- Create a circular pattern for the Lofted Blades.
- Enter **10** for number of instances.

5. Create a Revolve Cut:
- Use the Blade Trim Sketch and create a Revolve-Cut.

6. Save your work as:
- **Turbine Blades_Exe**.

Exercise: Advanced Sweep – Candle Holder

1. Opening the main sketch:

- From the Training Files folder open the part document named: **Candle Holder Sketch**.

- **Edit** this sketch and verify that the sketch is fully defined.

2. Revolving a surface:

- Click **Revolve Surface** .

- Use **Blind** and **360 deg**.

- Click **OK**.

3. Creating a Helix:

- Select the <u>Top</u> plane and open a new sketch.

- Sketch a **10.00in circle** centered on the origin.

- Convert the circle into a Helix using the settings as shown in the dialog box.

- **Exit** the Sketch.

4. Creating a swept surface:

- Select the <u>Top</u> plane once again and open another sketch.

- Sketch a Line from the Origin and **Pierce** the other end of the line to the Helix.

- Click and sweep the Line along the Helix using the **Swept-Surface** option.

- Click **OK**.

Pierce Relation

5. Create a new Axis: (to be used in step 9)

- Select **Insert / Reference Geometry / Axis** .

- Click the **Two-Planes** option.

- Select the **Front** and the **Right** planes from the FeatureManager tree.

- A preview of the new axis appears in the center of the part.

- Click **OK**.

6. Create the Intersection Curve:

- Hold the **Control** key and select the **Surface-Revolve1** and the **Surface Sweep1** from the FeatureManager tree.

- Click 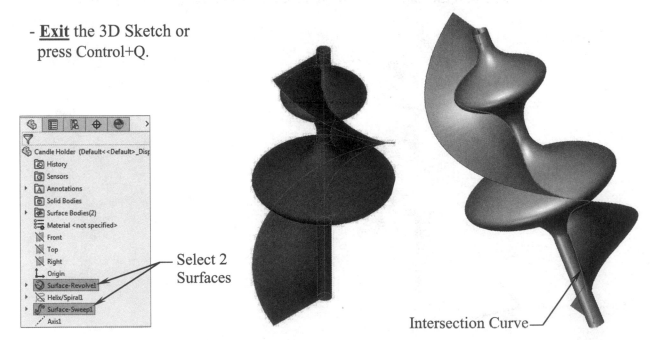 or select **Tools / Sketch Tools / Intersection Curve**.

- A 3D-Sketch is created from the intersection of the two surfaces.

- **Exit** the 3D Sketch or press Control+Q.

Select 2 Surfaces

Intersection Curve

7. Hide the 2 Surface Bodies:

- From the FeatureManager tree, right click on each Surface and select **Hide**.

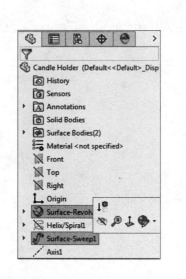

- This 3D sketch will be used as the sweep path in the next step.

- Click **OK**.

8. Create a solid swept feature:

- Switch to the Features tool tab and click or select **Insert / Boss-Base/ Sweep**.

- Select the **Circle Profile** option and enter **.260in** for Profile Diameter (arrow).

- Select the **3D-Sketch** for Sweep Path.

- Click **OK**.

9. Create the Circular Pattern :

- Using the **Axis** created in step 5 as the Center of the Pattern, repeat the Swept Feature **12** times.

- Click **OK**.

10. **OPTIONAL**:

- Create the **Base** and the **Candle Holder** solid features as shown below.

- Modify or design your own shapes if needed.

.125

.875

1.000

1.625

0

10.00°

R.375

22.750

22.625

22.500

R.125

21.125

20.250

0 (Zero is at
the Origin).

R.375

11. Save your work:

- Select **File / Save As**.

- Enter **Candle Holder** for file name.

- Click **Save**.

Advanced: Final Exam

- Create the part Bottle using the LOFT and SWEEP options where noted.

- All sketch profiles must be fully defined.

- The part must have no errors when finished.

1. Creating an Offset plane:

- Use the <u>Top</u> reference plane, create a new plane at **4.000 in.** offset distance.

2. Sketching the bottom profile:

- Select the <u>Top</u> plane and sketch the profile as shown.

- Use the Mirror option to create the **Symmetric** relations between entities.

- **<u>Exit</u>** the Sketch.

3. Creating the top profile:

- Use the <u>Plane1</u>; sketch a **Circle** centered on the Origin.

- Add a diameter dimension to fully define the sketch.

- **Exit** the sketch.

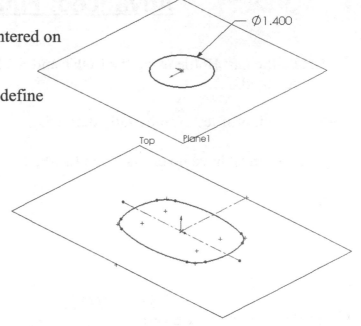

4. Creating the 1st Guide Curve:

- Select the <u>Front</u> plane and sketch the profile as shown.

* <u>Note:</u> *The construction lines will be used later to create the Derived-Sketch and the Guide Curves.*

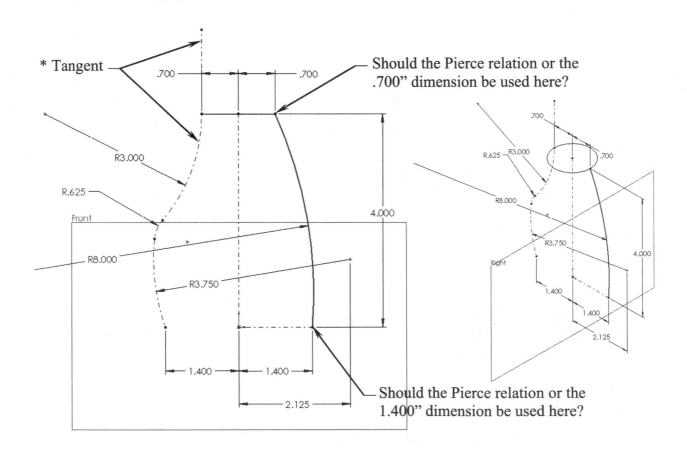

* Tangent

Should the Pierce relation or the .700" dimension be used here?

Should the Pierce relation or the 1.400" dimension be used here?

5. Creating the 2nd Guide Curve:

- Use the <u>Right</u> plane, sketch a **3-Point-Arc** and add dimensions.

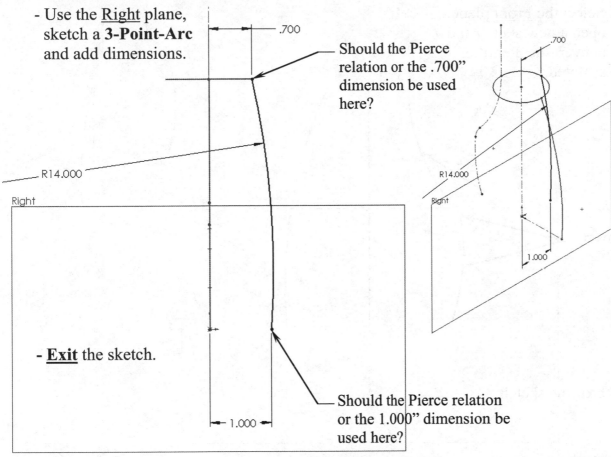

.700

Should the Pierce relation or the .700" dimension be used here?

R14.000

Right

1.000

- **Exit** the sketch.

Should the Pierce relation or the 1.000" dimension be used here?

6. Creating the 3rd Guide Curve:

- Use the <u>Right</u> plane, sketch a **3-Point-Arc** and add dimensions.

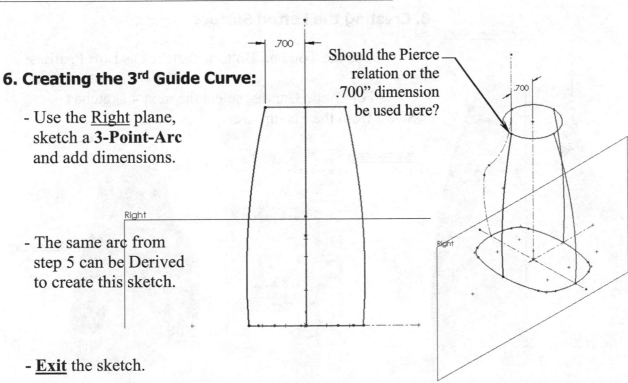

.700

Should the Pierce relation or the .700" dimension be used here?

Right

.700

Right

- The same arc from step 5 can be Derived to create this sketch.

- **Exit** the sketch.

7. Creating the 4th Guide Curve:

- Select the <u>Front</u> plane, open a new sketch and convert the 3 entities as noted.

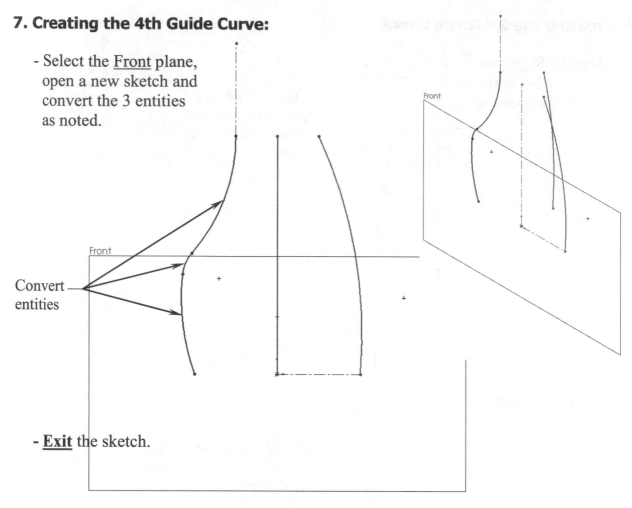

Convert entities

- <u>Exit</u> the sketch.

8. Creating the Lofted Surface:

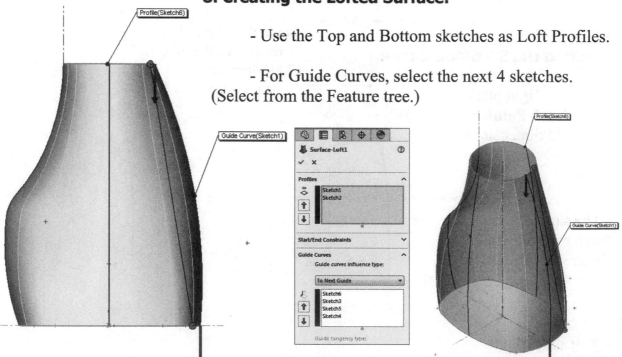

- Use the Top and Bottom sketches as Loft Profiles.

- For Guide Curves, select the next 4 sketches. (Select from the Feature tree.)

9. Filling the bottom surface:

- Select **Insert / Surface / Planar**.

    - Select **all edges** on the bottom for this operation.

- When finished, the bottom surface should be completely covered.

10. Sketching the Neck profile:

- Select the <u>Front</u> plane and sketch the profile below (2 lines).

- Revolve the sketch profile as a **Surface**.

- Revolve **One Direction**.

- Revolve a complete **360°**.

11. Knitting all surfaces into one:

- Select **Insert** / **Surface** / **Knit**.

- Select all **three surfaces**: the body, the Neck, and the Bottom surface.

- Clear the Gap Control option.

- When finished all 3 surfaces should now be combined as one continuous surface.

12. Adding fillet to the bottom edges:

- Add a **.250 in**. fillet to the bottom edges as shown.

Tips:

- *Right click on one of the edges and pick Select Tangency; this is one of the faster ways to select all edges at the same time.*

13. Adding fillet to the upper area:

- Add a **.093 in**. fillet to the upper edge as indicated.

Fillet 1 edge

14. Thickening the surface body:

- Select **Insert / Boss-Base / Thicken**.

- Enter a wall thickness of **.080 in**. to the **INSIDE** of the bottle.

15. Creating a new Offset plane:

- Select the Top face of the neck and create an Offset Distance plane at **.100 in.** below it.

16. Creating the Sweep Path of the Thread (the Helix):

- Convert the upper **circular edge** and then click **Insert / Curves / Helix-Spiral**.

- Pitch = **.125 in**.

- Revolution = **2.5**.

- Starting Angle = **0 deg**.

- Click **OK**.

Convert
Entity

17. Creating the Sweep Profile of the Thread:

- Select the <u>Right</u> plane and sketch the thread profile as shown.

- Add Dimensions and Relations needed to fully define the sketch.

- Add the **R.015"** fillets after the sketch is fully defined.

Pierce
relation

- **<u>Exit</u>** the sketch when finished.

18. Sweeping:

- Sweep the thread profile along the path to create the external threads.

19. Revolving:

- Convert the faces at the end of the thread and revolve them about the vertical centerlines to round off the ends.

20. Applying dimension changes:

- Change the dimension **R1.400** in the <u>Sketch1</u> to **R1.500**.

- Change the **Ø1.400** in the <u>Sketch2</u> to **Ø1.500**.

- Repair any errors caused by the changes.

21. Saving your work:

- Save your work as: **Level 3 – Final Exam**.

Designed by a CSWP student

Designed by a CSWP student

Designed by a CSWP student

Designed by a CSWP student

CHAPTER 12

SimulationXpress

SimulationXpress

SimulationXpress is a design analysis technology that allows the SOLIDWORKS users to perform first-pass stress analysis. SimulationXpress can help you reduce cost and time-to-market by testing your 3D designs within the SOLIDWORKS program, instead of expensive and time-consuming field tests.

There are five basic steps to complete the analysis using SimulationXpress:

1. Apply restraints (Fixture)
Users can define restraints. Each restraint can contain multiple faces. The restrained faces are constrained in all directions due to rigid body motion; you must at least restrain one face of the part to avoid analysis failure.

2. Apply loads
User inputs force and pressure loads to the faces of the model.

3. Define material of the part
* EX (Modulus of elasticity).
* NUXY (Poisson's ratio). If users do not define NUXY, SimulationXpress assumes a value of 0.
* SIGYLD (Yield Strength). Used only to calculate the factors of safety (FOS).
* DENS (Mass density). Used only to include mass properties of the part in the report file.

4. Analyze the part
SimulationXpress prepares the model for analysis, then calculates displacements, strains, and stresses.

5. View the results
After completing the analysis, users can view results. A check mark on the Results tab indicates that results exist and are available to view for the current geometry, material, restraints, and loads. A report can also be created in MS-Word format.

SimulationXpress
Using the Analysis Wizard

von Mises (psi)

- 6.095e+004
- 5.587e+004
- 5.079e+004
- 4.572e+004
- 4.064e+004
- 3.556e+004
- 3.049e+004
- 2.541e+004
- 2.033e+004
- 1.525e+004
- 1.018e+004
- 5.101e+003
- 2.476e+001

View Orientation Hot Keys:

Ctrl + 1 = Front View
Ctrl + 2 = Back View
Ctrl + 3 = Left View
Ctrl + 4 = Right View
Ctrl + 5 = Top View
Ctrl + 6 = Bottom View
Ctrl + 7 = Isometric View
Ctrl + 8 = Normal To
       Selection

Dimensioning Standards: **ANSI**

Units: **INCHES** – 3 Decimals

Tools Needed:

SimulationXpress is part of
SOLIDWORKS 2019 Basic,
SOLIDWORKS Office Professional,
and SOLIDWORKS Premium.

1. Starting SimulationXpress:

- SimulationXpress is an
introductory version of
fully featured simulation.

- SimulationXpress is a design analysis application that is fully integrated with
SOLIDWORKS. It is used by designers, analysts, engineers, students, and others
worldwide to design safe, efficient, and economical products.

- To enable SimulationXpress, visit www.my.solidworks.com/xpress, log in or
create a user account and enter your SOLIDWORKS Serial Number to generate
the Xpress Product Codes. You can find your serial number from the **Help** menu
> **About SOLIDWORKS**.

- Select **Tools / Xpress Products / SimulationXpress**. After entering the Xpress
Product Code, click **OK** to launch the SimulationXpress application.

- The SimulationXpress program is launched and appeared on the right side of the
screen.

2. Opening a part document:

- Open the part document that was created earlier: **Spanner** (or open a copy from the Training Files folder).

- **Suppress** the extruded text (the 5/8" and the Spanner text).

- From the **Tools** drop-down menu, select **SimulationXpress** (Arrow).

Suppress the text
(top & bottom)

3. Setting up the Units:

- Click **Options** (arrow)
to set the system of
units for the analysis.

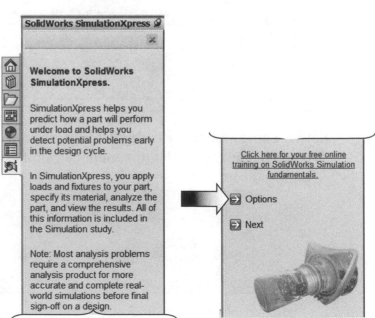

- Select **English (IPS)** for System Of Units (Inch, Pound, Second).

- Select the folder and the location to save the analysis results.

- Enable the option:

Show Annotation for Maximum and Minimum in the Result Plot.

- Creating a new folder for each study is recommended.

- Click **OK** .

- Click **Next** →.

4. Adding a Fixture (restraint):

- The next step is to create the restraint area(s).

- Each restraint can contain one or multiple faces. The restrained faces are constrained in all directions. There must be at least one fixed face of the part to avoid analysis failure due to rigid body motion.

> 🔆 **Restraints**
>
> Restraint is used to anchor certain areas of the model so that they will not move or shift during the analysis. At least one face should be restrained prior to running the analysis.

- Click **Add a Fixture** →.

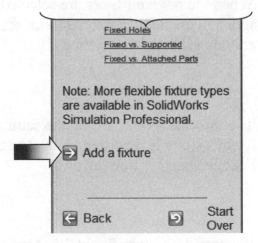

- Select the 3 faces as indicated to use as restraint faces.

- The Restraint faces are locked in all directions to avoid failure due to rigid body motion.

Restraint
face 1…

Restraint
face 2 & 3

- Click **OK**.

- When the restraint faces are selected, more faces can be added to create different restraint sets. They can also be edited or deleted at anytime.

- The information regarding the settings and parameters for this study is recorded on the lower half of the FeatureManager tree (arrows).

- Rename the fixture Fixed-1 to **Open-End**.

5. Applying a Force:

- Click **Add a Force** →.

- The **Forces** and **Pressures** options allow SOLIDWORKS users to apply force or pressure loads to faces of the model. Multiple forces can be applied to a single face or to multiple faces.

- Select the 2 faces from the back side of the closed end as shown below.

Select 2 faces

Force & Load

* Force infers a total force applied to a face in a specific direction.

* Load infers that the force is evenly distributed on the face.

- Click the **Selected Direction** and select the **Front** plane from FeatureManager tree (arrow).

- Click the **Total** option (arrow). Enter **100** lbs. for Total Force value (arrow).

- Click Reverse Direction if needed.

- Click **OK**. Rename the Force-1 to **Closed-End**.

- At this point we will need to specify a material so that SimulationXpress can predict how it will respond to the loads.

6. Selecting the material:

- Click **Next** →.

- Select **Choose Material** →.

- Expand the **Steel** option (click the + symbol).

- Choose **Alloy Steel** from the list (arrow).

Material Editor

Material can be assigned to the part using the **Material Editor** PropertyManager. The material will then appear in SimulationXpress.

- The analysis results are dependent on the material selection. Simulation-Xpress needs to know the Visual and the Physical properties to run the analysis.

- Click **Apply** and **Close** .

- The material **Alloy Steel** is now assigned to the part. SimulationXpress assumes that the material deforms in a linear fashion with increased load. Non-Linear materials (such as many plastics) require the use of Simulation Premium.

- The Modulus of Elasticity and the Yield Strength for the selected material are reported on the right.

- Click **Next** .

- SimulationXpress is ready to analyze the model based on the information provided. Displacements, Strains, and Stresses will then be calculated.

7. Analyzing the model:

- Click **Run Simulation** .

- SimulationXpress automatically tries to mesh the model using the default element size. The smaller the element size the more accurate the results, but more time is needed to analyze the model.

- To change mesh settings: First click **Change settings** then click **Change mesh density**.

    * Drag the slider to the right for a finer mesh (more accurate, but takes longer).

    * Drag the slider to the left for a coarser mesh (quicker).

- Select **Yes, continue** .

8. Viewing the Results:

- Click **Show Von Mises Stress** .

- SimulationXpress plots stresses on the deformed shape of the part.

- In most cases, the actual deformation is so small that the deformed shape almost coincides with the un-deformed shape, if plotted to scale.

- SimulationXpress exaggerates the deformation to demonstrate it more clearly.

- The Deformation Scale shown on the stress and deformed shape plots is the scale used to rescale the maximum deformation to **4.34794%** of the bounding box of the part.

- The **Stress Distribution Plot** is displayed below.

- To see the resultant displacement plot click **Show Displacement** ➡.

- Click **Play Animation** ▶ or **Stop Animation** ⬜ when finished viewing.

Displacement Plot

- To view regions of the model with a factor of safety less than a given value (1), click: **Show Where Factor Of Safety (FOS) Is Below: 1** (or enter any value).

- SimulationXpress displays regions of the model with factors of safety less than the specified value in red (unsafe regions) and regions with higher factors of safety in blue (safe regions).

- When finished viewing click: **Done Viewing Results**.

9. Creating the report:

- Click **Generate Report** ➡.

- SimulationXpress cycles through the results, generates a report in Word format, and the MS Word program is launched to display the full report.

- The report includes:

1. Cover Page   2. Model Information

2. Load/Fixture Details   4. Slid Mesh Information

5. Stress Results   6. Displacement Results

7. Deformation Results   8. Factor of Safety Results

- In the Report Settings dialog box, enable the Description checkbox and enter the following:

* Spanner - SimXpress Study.

* Your Name.

* Your Company Name and any information that you wish to include in the report.

- Select the location to save the report.

- Click **Generate**.

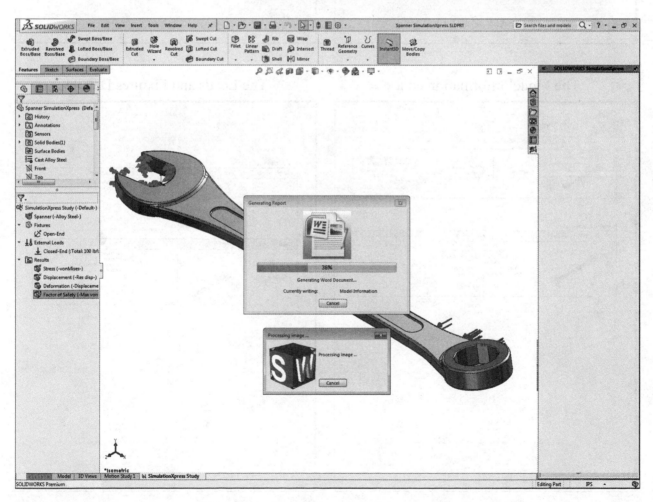

The Report Cover Page

The Model Information

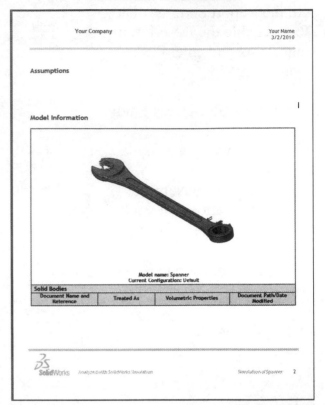

The Model Information cont.

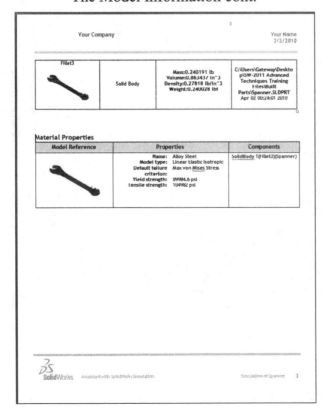

The Loads and Fixtures Details

The Mesh Information

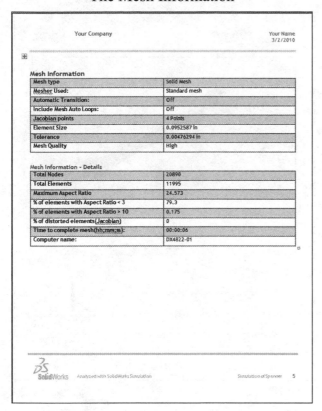

The Solid Mesh Plot

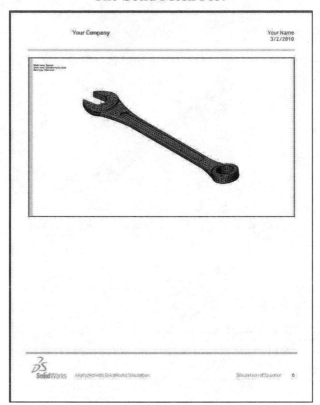

The Von Mises Stress Plot

The Displacement Plot

The Deformation Plot

The Factor of Safety Plot

 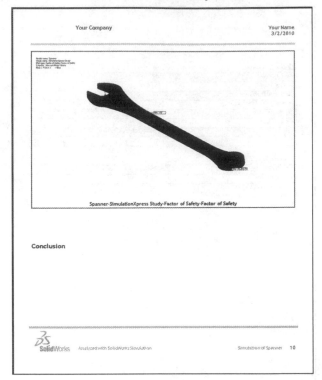

- When finished with viewing the Report, click: **Generate eDrawings File** →.

10. Generating the eDrawings file:

- An eDrawings file can be created for the SimulationXpress result plots. The eDrawings file allows you to view, animate and print your analysis results.

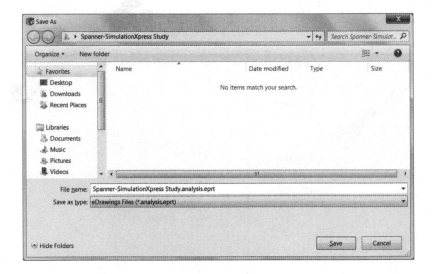

- When prompted, **Save** the analysis study in the default folder.

- SimulationXpress creates an eDrawing file with the **.eprt** extension. The file contains von Mises stress, displacement, deformation, and Factor of Safety plots. By default, the von Mises stress plot is displayed.

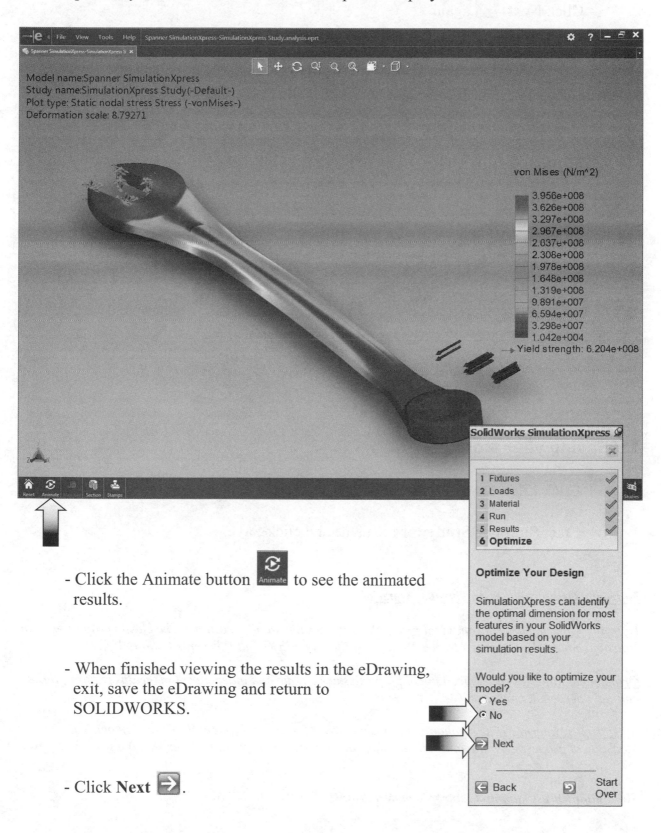

- Click the Animate button 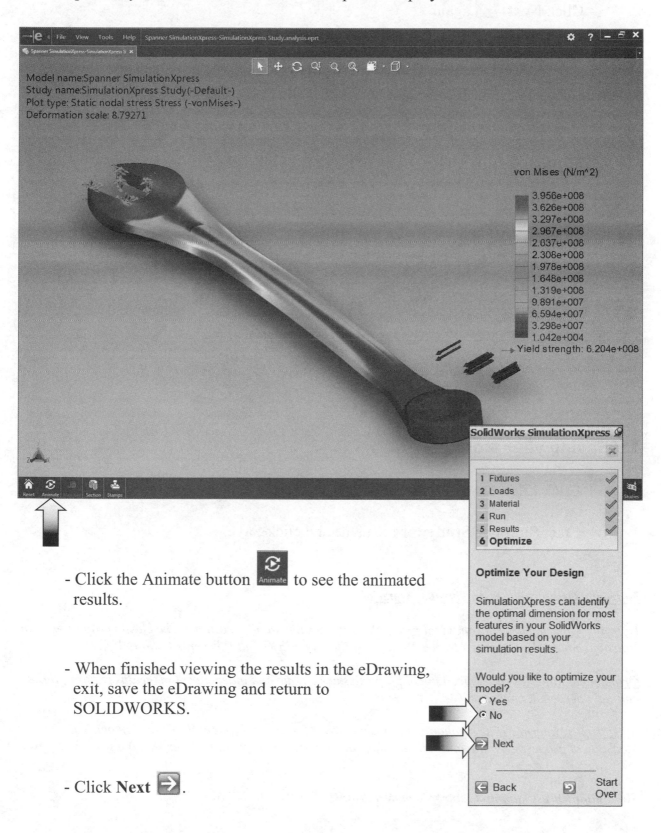 to see the animated results.

- When finished viewing the results in the eDrawing, exit, save the eDrawing and return to SOLIDWORKS.

- Click **Next** .

- Click **No** under **Optimize Your Design.** (Only if the analysis fails then the optimization step is needed.)

- Click **Next** → again.

- At this point, you are prompted that the analysis has been completed. All 5 steps on the SimulationXpress property tree (right side) have the check marks in front of them.

11. Saving your work:

- Click **File / Save As**.

- Enter **Spanner Study** for file name and click **Save**.

<u>Isotropic, Orthotropic & Anisotropic Materials:</u>

<u>Isotropic Material</u>: If its mechanical properties are the same in all directions. The elastic properties of an Isotropic material are defined by the Modulus of Elasticity (EX) and Poisson's Ratio (NUXY).

<u>Orthotropic Material</u>: If its mechanical properties are unique and independent in the directions of three mutually perpendicular axes.

<u>Anisotropic Material</u>: If its mechanical properties are different in different directions. In general, the Mechanical properties of the anisotropic materials are not symmetrical with respect to any plane or axis.

SimulationXpress supports Isotropic materials only.

Questions for Review

SimulationXpress

1. SimulationXpress can be accessed from the Tools pull down menu.
    a. True
    b. False

2. System Of Units SI (Joules) is the only type that is supported in SimulationXpress.
    a. True
    b. False

3. The material of the part can be selected from the built-in library or input directly by the user.
    a. True
    b. False

4. SimulationXpress supports Isotropic, Orthotropic, and Anisotropic materials.
    a. True
    b. False

5. Restraints/Fixture are used to anchor certain areas of the part so that it will not move during the analysis.
    a. True
    b. False

6. Only one surface/face should be used for restraint in each study.
    a. True
    b. False

7. The elements size (mesh) can be adjusted to a smaller value for more accurate results.
    a. True
    b. False

8. The types of results reported are:
    a. Stress Distribution
    b. Deformed Shape
    c. Deformation
    d. Factor of Safety
    e. All of the above

7. TRUE	8. E
5. TRUE	6. FALSE
3. TRUE	4. FALSE
1. TRUE	2. FALSE

Exercise 1: SimulationXpress: Force

1. Open the existing part: **Extrude Boss & Extrude Cut.**

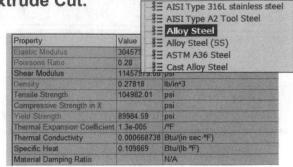

2. Set Unit to: **English (IPS).**

5. Select Material: **Alloy Steel.**

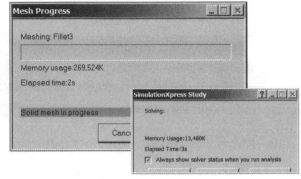

6. Run the Analysis.

3. Apply Restraint: **to 3 Holes.**

7. Check the von Mises Stress results.

4. Apply Force of **50 Lbs** to the side hole, normal to the **Front** Plane.

8. Save a copy as: **Simulation_Force.**

Exercise 2: SimulationXpress: Pressure

1. Open the existing part: **Bottle_SimulationXpress**.

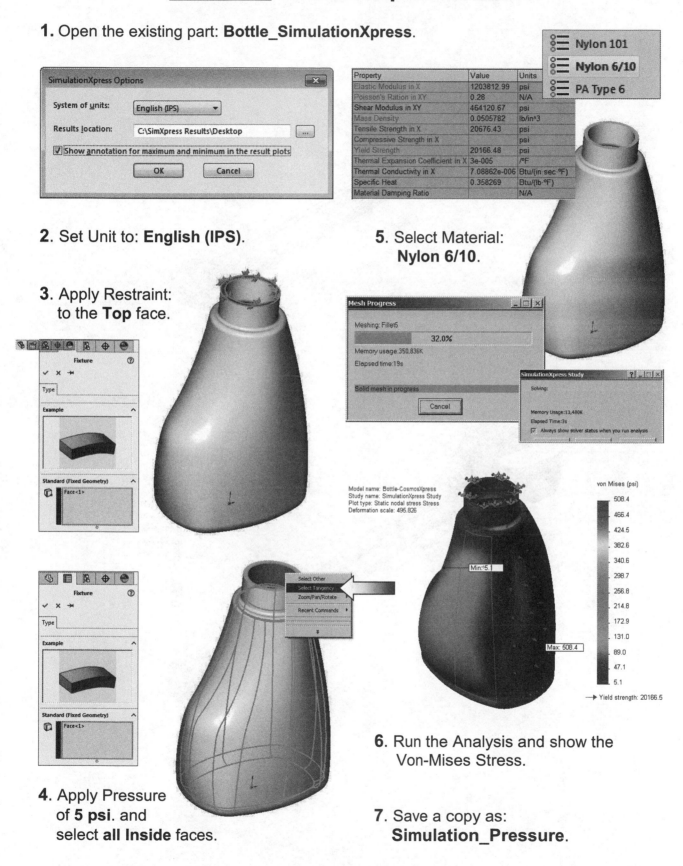

2. Set Unit to: **English (IPS)**.

3. Apply Restraint:
to the **Top** face.

4. Apply Pressure
of **5 psi**. and
select **all Inside** faces.

5. Select Material:
Nylon 6/10.

6. Run the Analysis and show the
Von-Mises Stress.

7. Save a copy as:
Simulation_Pressure.

von Mises (psi)
6.095e+004
5.587e+004
5.079e+004
4.572e+004
4.064e+004
3.556e+004
3.049e+004
2.541e+004
2.033e+004
1.525e+004
1.018e+004
5.101e+003
2.476e+001

CHAPTER 13

Sheet Metal Parts

Sheet Metal Parts

This chapter discusses the introduction to designing sheet metal parts.

Create a sheet metal part in the folded stage and add the sheet metal specific flange features such as:

* Base Flange.
* Edge Flanges.
* Sketch Bends.
* Cut with Link to Thickness.
* Normal Cuts.

There are at least 3 options for specifying the setback allowance, or the length difference between the fold and the flat patterns of a sheet metal part.

* **Bend Table:** You can specify the bend allowance or bend deduction values for a sheet metal part in a bend table. The bend table also contains values for bend radius, bend angle, and part thickness.

* **K-Factor:** Is a ratio that represents the location of the neutral sheet with respect to the thickness of the sheet metal part.
  Bend allowance using a K-Factor is calculated as follows:
  $$BA = \Pi(R + KT) \; A/180$$

* **Use Bend Allowance:** Enter your own bend value based on your shop experience.

* **Bend allowance Calculations:** The following equation is used to determine the total flat length when bend allowance values are used: $L_t = A + B + BA$

* **Bend Deduction Calculations:** The following equation is used to determine the total flat length when bend deduction values are used: $L_t = A + B - BD$

Sheet Metal Parts
Post Cap

View Orientation Hot Keys:

Ctrl + 1 = Front View
Ctrl + 2 = Back View
Ctrl + 3 = Left View
Ctrl + 4 = Right View
Ctrl + 5 = Top View
Ctrl + 6 = Bottom View
Ctrl + 7 = Isometric View
Ctrl + 8 = Normal To
        Selection

Dimensioning Standards: **ANSI**

Units: **INCHES** – 3 Decimals

Tools Needed:

 Insert Sketch   Line   Circle

 Dimension   Add Geometric Relations   Base Flange

 Edge Flange   Sketch Bend   Flat Pattern

1. Starting with the base profile:

- Select <u>Right</u> plane from FeatureManager tree and insert a new sketch [icon].

- Sketch the profile below using the Line tool [icon].

- Add dimensions [icon] and relations [icon] needed; fully define the sketch.

4.000

If Mid-Point relation is used, remove the dim 2.000

4.000

2.000

4.000

Gauge Tables

To enable the Gauge Tables while in the Base Flange mode, click: **Browse / Installed Dir. / SolidWorks / Lang / English / Sheet Metal Gauge Tables**.

---

2. Extruding the Base Flange:

- Click [icon] or select **Insert / Sheet Metal / Base Flange**.

- Direction 1:   **Mid-Plane**.

- Extrude Depth: **4.00 in**.

- Use Gauge Table: **Sample Table - Steel**.

- Mat'l Thickness: **16 Gauge (.0598)**.

- Bend Radius: **.030 in**. (Override Radius).

Set material thickness to **Outside**.

- Click **OK**.

3. Creating an Edge Flange:

- Hold the **Control** key and select the **2 edges** as indicated.

- Click or select **Insert / Sheet Metal / Edge Flange**.

——— Select 2 edges

- Drag the cursor downwards and click anywhere to lock the preview of the 2 Edge Flanges.

- Select **Material Outside** under the Flange Position.

- Enter **90°** for Angle (default).

- Flange Length: **Blind**.

- Enter **5.000"** for depth.

- Relief Type: **Rectangle**.

- Relief Ratio: **.500**.

- Click **Edit Flange Profile** (arrow).

** The flange length will be modified
in the next steps.*

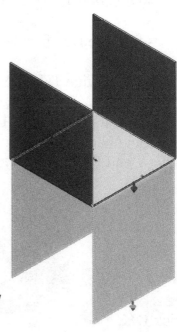

4. Editing the Edge Flange Profile:

- Move the Profile Sketch dialog out of the way.

- Drag the 2 outer lines inward as noted.

- Add the dimensions shown to fully define the sketch.

- After the 1st edge flange becomes fully defined, click the **Back** button (arrow) to switch back to the previous screen.

- Select the **Edge 2** under the Flange Parameters section and click the **Edit Flange Profile** once again (arrow).

- Drag the 2 vertical lines inward similar to step above, and add the same dimensions as the 1st flange to ensure the 2 flanges are exactly the same size (Collinear relations can also be used for the width and height of the 2nd flange).

- Click finished 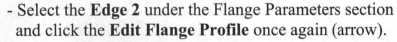 after the 2nd sketch becomes fully defined.

- Zoom in on the upper end of the edge flange to see the relief details.

- Beside the Rectangular relief, there are two other types available: Obround and Tear. Select the **Rectangular** type.

- Auto-Relief is added automatically after exiting the Sketch.
- Relief Width and Depth are defaulted to one-half (0.5) the Material Thickness.

5. Viewing the Flat Pattern:

- Click **Flatten** on the Sheet Metal toolbar:

- The flat pattern of the part is displayed.

- Another way to view the flat pattern is to right click on the Flat-Pattern1 feature and select **Suppress** (arrow).

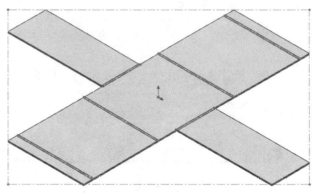

6. Changing the Fixed face:

- Edit the **Flat-Pattern** feature and select the <u>face</u> indicated to use as the Fixed face.

Select the Fixed face

7. Creating a Sketch Bend:

- Select the **side face** and open a new Sketch 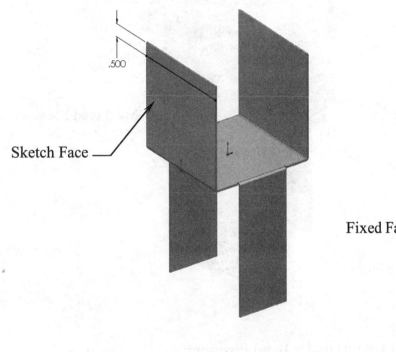.

- Sketch a **line** starting at one edge and Coincident with the other.

- Add dimensions to fully define the sketch.

.500

Sketch Face

.500

Fixed Face

- Click 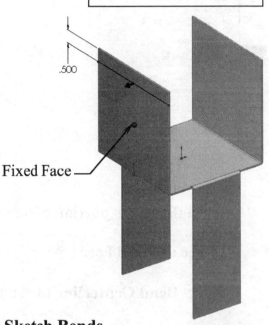 or select **Insert / Sheet Metal / Sketch Bends**.

- Select the **lower portion** of the surface as Fixed Face.

- Select **Bend Centerline** (default) under Bend Position.

- Enter **90.00** deg for Bend Angle.

- Enable **Use Default Radius**.

- Click **OK**.

8. Creating the 2ⁿᵈ Sketch Bend:

- Select the surface on the right and open a sketch ⌐ or select **Insert / Sketch**.

- Sketch a line ⟋ as shown and add dimensions ⟍ to fully define.

- Click 🖨 or select **Insert / Sheet Metal / Sketch Bend**.

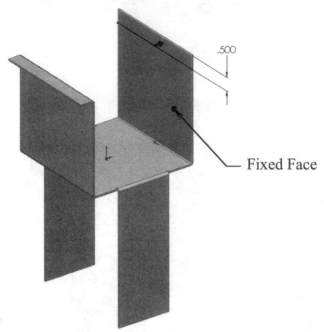

.500

Fixed Face

- Select the **lower portion** of the

surface as Fixed Face .

- Select **Bend Centerline** (default) under Bend Position.

- Enter **90.00** deg for Bend Angle (default).

- Enable **Use Default Radius**.

- Click **OK**.

- The resulting bend.

- The upper portion of the flange is bent outward 90°,
leaving the lower portion fixed.

9. Adding holes on the Edge Flange:

- Open a new sketch on the side <u>face</u> as indicated 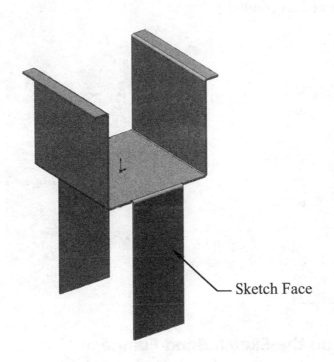 or select:
  Insert / Sketch.

Sketch Face

- Sketch two circles on the face.

- Add dimensions as shown to position the circles.

- Add **Vertical** relations between the centers of the circles and the Origin, to fully define the sketch.

1.500

2.500

Vertical

⌀.225

- Click or select **Insert / Cut / Extrude**.

- End Condition: **Through All**.

- Enable **Normal Cut** (default).

- Click **OK** ✅.

10. Adding holes on the Sketch Bend Flanges:

- Select the face as noted and open a new sketch 🗔 or select **Insert / Sketch**.

- Sketch 4 circles ⊙ and add dimensions 🗇 as shown.

- Add the relations ⊥ as needed to fully define the sketch.

- Click 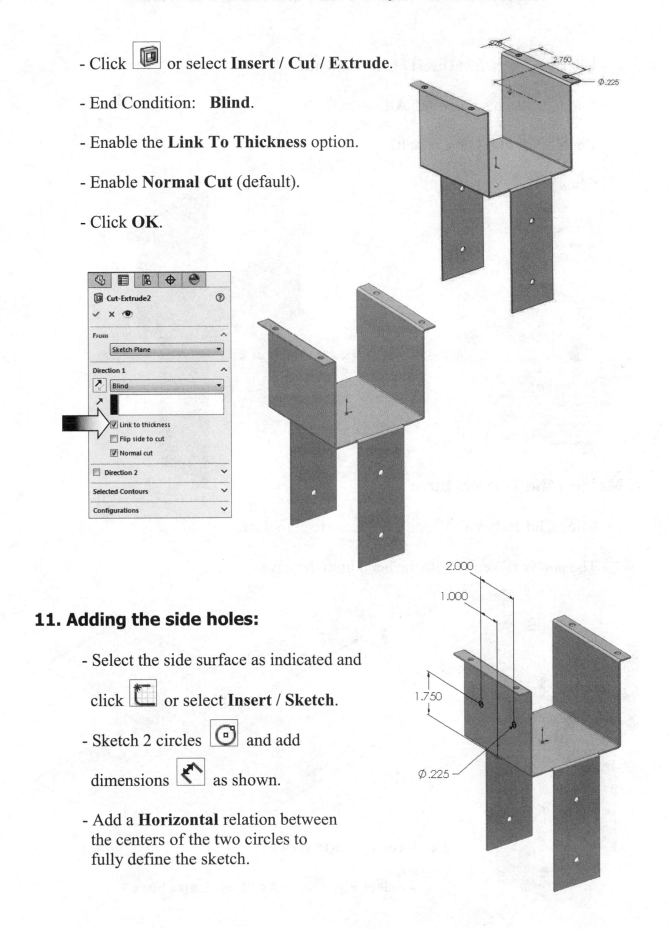 or select **Insert / Cut / Extrude**.

- End Condition: **Blind**.

- Enable the **Link To Thickness** option.

- Enable **Normal Cut** (default).

- Click **OK**.

11. Adding the side holes:

- Select the side surface as indicated and

  click or select **Insert / Sketch**.

- Sketch 2 circles and add

  dimensions as shown.

- Add a **Horizontal** relation between
  the centers of the two circles to
  fully define the sketch.

- Click or select **Insert / Cut / Extrude**.

- End Condition: **Through All**.

- Enable **Normal Cut** (default).

- Click **OK**.

12. Making the Flat Pattern:

- Click **Flat Pattern** on the Sheet Metal toolbar.

- The part is flattened with the bend lines displayed.

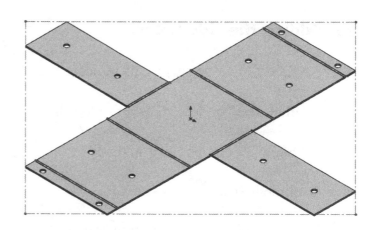

13. Saving your work:

- Select **File / Save As / Post Cap / Save**.

Using Sheet Metal Costing

Use SOLIDWORKS sheet metal Costing tools to determine the cost of a sheet metal. Costing provides a comprehensive breakdown and comparison of manufacturing and material costs for sheet metal parts.

Costing Template Editor for Sheet Metal Parts

You can create and edit costing templates for sheet metal parts or bodies from the Costing Template Editor.
You can specify rates and costs for the procedures required to manufacture a sheet metal part or body in the sheet metal template. You can include customized information in the template, such as material cost and thicknesses, cost of manufacturing, and manufacturing setup costs. You can determine how manufacturing operations affect the cost of your design. For example, you can set up templates for vendors that use different manufacturing operations.

You can specify the file location for Costing templates in Tools > Options > System Options > File Locations. In Show folders for, select Costing templates to add or delete a location. The default Costing template folder is *install_dir/lang/language/* Costing templates.

1. Opening an existing sheet metal part:

- Click **File / Open**.

- Browse to the Training Files folder, locate and open the **Sheet Metal Costing** part document.

- The default template and all its values will be used in this exercise.

2. Inputting the information:

- Switch to the **Evaluate** tab and click **Costing** 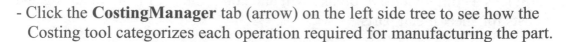 .

- Click the **CostingManager** tab (arrow) on the left side tree to see how the Costing tool categorizes each operation required for manufacturing the part.

- Select the **Material** to **Steel AISI 304)** and **Thickness** to **1.90mm** (.075in). The sheet metal part is evaluated based on the selected material and processing costs as set in the default template.

- The Costing tab (on the right) indicates the cost for the selected material and thickness is roughly 16.09 USD per kg.

- The total cost per part is **24.12 USD**.

- The major processing cost (on the left) is **$2.16 USD** for 9 laser cut paths and **$8.00 USD** for eight bends ($1.00 USD per bend).

3. Setting the Baseline:

- The **Set Baseline** is used to set a baseline cost for comparison. If you change the design later on, the cost is compared to the baseline cost. When you set a baseline cost, any changes to the part are considered Current and the difference is displayed. While the baseline price is set, the part is rotated, flattened, and refolded because the software is capturing images for the Costing report.

- Click the **Set Baseline** button .

- The final cost is based on the values for manufacturing steps. The values that we are using here are only for samples only.

- Change the material to:
  Aluminum Alloy 6061

- Use the same thickness (1.90mm or .075in).

- The SOLIDWORKS Costing recalculates the cost based on the change in material.

- The recalculated cost is now almost half of what it was for Steel ($14.69 instead of $24.12).

4. Saving your work:

- Click **File / Save As**.

- Enter **Sheet Metal Costing** for the name of the document and click **Save.**

- **<u>General</u>**
  Use the General screen in the Costing Template Editor to set the units and currency options.

- **<u>Material</u>**
  Use the Material screen in the Costing Template Editor to set the materials you need to manufacture the sheet metal part.

- **<u>Thickness</u>**
  Use the Thickness screen in the Costing Template Editor to set the thickness and cost values for each class and material combination.

- **<u>Cut</u>**
  Use the Cut screen in the Costing Template Editor to define the cost of cutting methods based on length or stroke.

- **<u>Bend</u>**
  Use the Bend screen in the Costing Template Editor to define the cost of bending methods based on regular bends or hem bends.

- **<u>Library Features</u>**
  Use the Library Features screen in the Costing Template Editor to define the cost of library features, punch features, and forming tools in the part.

- **<u>Custom</u>**
  Use the Custom screen in the Costing Template Editor to define additional operations that contribute to a part's manufacturing cost, such as powder coating.

Questions for Review

Sheet Metal Parts

1. A sheet metal part can have multiple thicknesses.
   a. True
   b. False

2. An Edge Flange can be mirrored just like any other feature.
   a. True
   b. False

3. A sheet metal part can be created right from the beginning using the Base flange option.
   a. True
   b. False

4. Auto relief option is not available when extruding a Base Flange.
   a. True
   b. False

5. When the Sketched Bend option is used to create a bend, you'll have to specify at least two parameters:
   a. A fixed side and a sketched line.
   b. Fixed side and a bend angle value.
   c. A bend radius and a bend angle value.

6. The only time when the K-Factor option can be changed to Bend Table is when extruding the Base Flange.
   a. True
   b. False

7. A sheet metal part can be designed from a flat sheet and other bends can be added later using Sketched Bend, Edge Flange, etc.
   a. True
   b. False

8. Link-to-Thickness option allows all sheet metal features in a part to have the same wall thickness and they can all be changed at the same time.
   a. True
   b. False

7. TRUE	8. TRUE
5. A	6. FALSE
3. TRUE	4. FALSE
1. FALSE	2. TRUE

Sheet Metal Parts

Sheet Metal Parts

Sheet metal parts can be created using one of the following methods:

- Create the part as a solid and then insert the sheet metal parameters such as rips, bend radius, material thickness, bend allowance, and cut relief so that the part can be flattened.

- Create the part as a sheet metal part from the beginning by using the Base Flange command to extrude the first feature.

- Sheet metal parameters can be applied onto the sheet metal part during the extrusion or after the fact.

This chapter will guide you through the use of the sheet metal and forming tool commands to create a sheet metal part and a louver form tool:

- Creating the parent feature with the Base Flange command.
- Using the Miter-flange command.
- Create a sheet metal part in the flat or folded stage.
- Create Revolved features.
- Accessing the Design Library.
- Using the forming tool to form the louvers.
- Create a Linear pattern of features.
- Create a Circular pattern of features.
- Create a pattern of patterned features.

Sheet Metal Parts
Vents

Dimensioning Standards: **ANSI**

Units: **INCHES** – 3 Decimals

Tools Needed:

Insert Sketch	Line	Dimension
Rectangle	Add Geometric Relations	Linear Pattern
Base Flange	Miter Flange	Circular Pattern
Flat Pattern	Extruded Cut	Design Library

1. Sketching the first profile:

- Select <u>Front</u> plane from the FeatureManager tree.

- Click from Sketch toolbar, or select **Insert / Sketch**.

- Click and sketch a horizontal line **below the Origin**.

- Click and dimension the length of the line to **18.00 in**.

—Origin

2. Adding a Midpoint relation:

- Click from the Sketch toolbar OR select **Tools / Relations /Add**.

- Click on the Origin point and select the line as shown.

- Select **Midpoint** option from the Add Geometric Relation dialog box.

- Click **OK**.

Mid-Point relation

3. Extruding the Base-Flange:

- Click 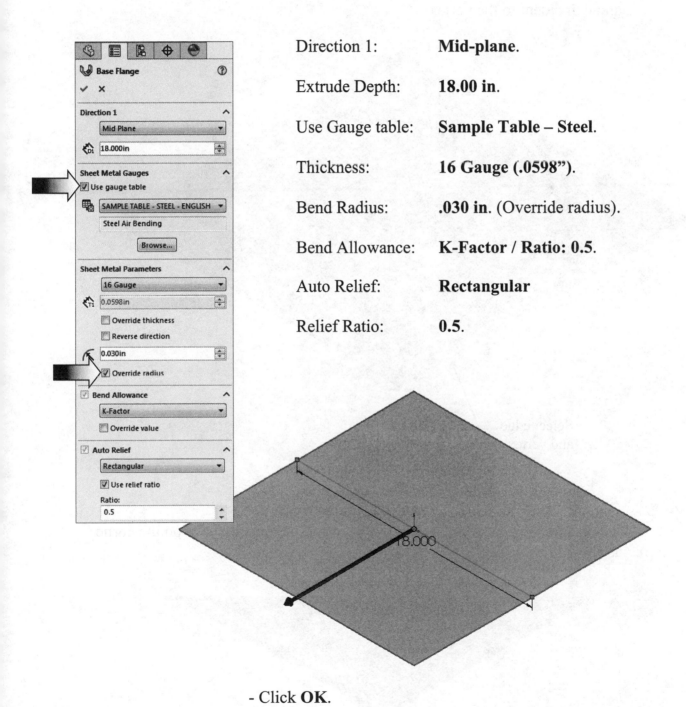 (Base Flange) from the Sheet Metal toolbar or select:
Insert /Sheet Metal / Base Flange.

- Enter / select the following:

Direction 1:	**Mid-plane**.
Extrude Depth:	**18.00 in**.
Use Gauge table:	**Sample Table – Steel**.
Thickness:	**16 Gauge (.0598")**.
Bend Radius:	**.030 in**. (Override radius).
Bend Allowance:	**K-Factor / Ratio: 0.5**.
Auto Relief:	**Rectangular**
Relief Ratio:	**0.5**.

- Click **OK**.

4. Creating the Miter-Flanges:

- Hold the control key, select the <u>edge</u> and the <u>vertex</u> as indicated below, and

  click 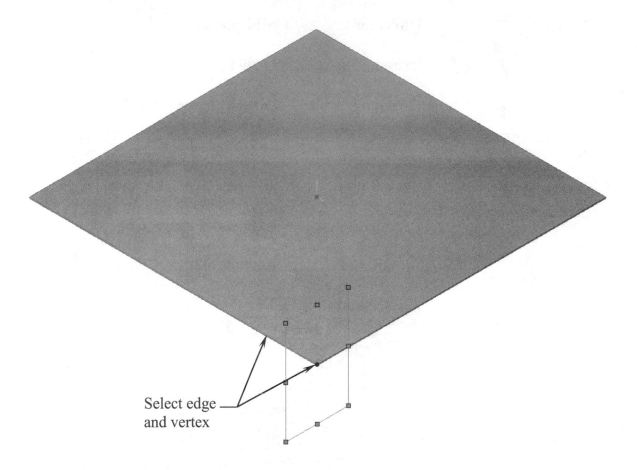 or select: **Insert / Sketch**.

- SOLIDWORKS automatically creates a new plane <u>normal</u> to the selected edge and coincident to the vertex.

Select edge and vertex

- Click **Zoom-to-Area** from the View toolbar and zoom in on the corner as shown below.

- Sketch a vertical line starting at the upper corner.

- Click (Dimension) and make the length of the line **.500 in**.

.500

- Click (**Miter-Flange**) from the Sheet Metal toolbar or select
Insert / Sheet Metal / Miter Flange.

- Select the **4** upper edges as indicated.

- Choose **Material Outside** under Flange Position.

- Set Gap Distance to **.010 in**.

- Click **OK**.

Select the top
4 edges

5. Flattening the part:

- The Flat Pattern can be toggled at anytime during or after the part is created.

- Click (Flatten) from the Sheet Metal toolbar to flatten the part.

6. Switching back to the folded-model:

- Click (Flatten) again to return it back to the folded stage.

7. Saving your work: Save your model as **Sheet Metal Vent.sldprt**

---

8. Creating a new Forming Tool – The Louver:

- Start a new Part file: click **File / New / Part / OK.**

- Set Units to Inches – 3 Decimal (Tools/Options/Document Properties/ Units).

9. Sketching on the TOP reference plane:

- Select the <u>Top</u> plane and click **Insert / Sketch** from the Sketch toolbar.

- Select the **Center Rectangle** from the Sketch-Tools toolbar.

- Sketch a rectangle that is centered on the Origin, as shown below.

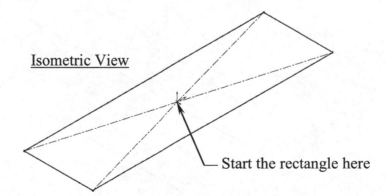

<u>Top View</u>

<u>Isometric View</u>

Start the rectangle here

10. Other Rectangle options:

 <u>**Corner Rectangle**</u>

Click 2 diagonal points.

 <u>**3-Point Corner Rectangle**</u>

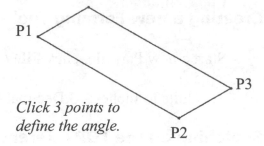

Click 3 points to define the angle.

 <u>**3-PointCenter Rectangle**</u>

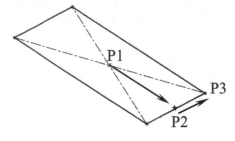

Start at Center point and click 2 other points to define the angle.

 <u>**Parallelogram**</u>

Click 3 points to define the parallelogram.

11. Adding dimensions:

- Click (Smart Dimension) and add the dimensions shown below.

8.000

2.000

12. Extruding the Base:

- Click (**Extruded Boss/Base**) and fill in the following parameters:

- End Condition: **Blind**.

- Depth: **.125 in**.

- Click **OK**.

> **Forming Tools**
>
> Forming tools are solid parts that are used to bend, stretch, and form sheet metal.

13. Building the louver body:

- Select the <u>upper face</u> of the part and open a new sketch.

- Change to the Top View Orientation (Ctrl+5).

- Sketch the profile of the Louver-Forming tool and add dimensions as shown below:

- Select the horizontal centerline and click (**Revolve Boss/Base**).

- For Revolve Type, select **Blind**.

- For revolve Angle, enter **90°**.

- Toggle (Reverse) and make sure the preview looks like the one below.

- Click **OK**.

- The resulting 90° revolved.

14. Adding a fillet at the base:

- Click **Fillet** and enter **.125 in**. for Radius size.

- Select the **edge** as indicated.

- Select **Tangent Propagation** checkbox (default); the system applies the same fillet to all connected tangent edges.

- Click **OK**.

- Rotate the model to see the resulting fillet from the backside.

- The new fillet should run around the back side but not the front.

15. Removing the base:

- The rectangular block was created
  so that a fillet can be added between
  the two features. We no longer
  need it at this point.

Sketch face

- Select the <u>side surface</u> of the part
  and open a new sketch.

- While the side surface is still highlighting
  press the **Convert Entities** command.
  The selected face is converted to a
  rectangle.

- Switch to the **Features** tool tab
  and click the **Extruded Cut** command.

- Change the End Condition to **Through All**.

- Click **OK**.

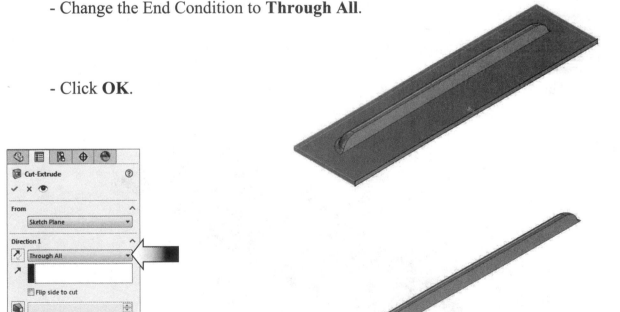

- The base is removed and
  only the form tool portion
  is kept.

16. Creating the Positioning Sketch:

- The Positioning Sketch displays the preview of the Form tool while it is being dragged from the Design Library. Its sketched entities can be dimensioned to position the Formed feature.

- Select the **bottom face** of the part and open a new sketch [⬚].

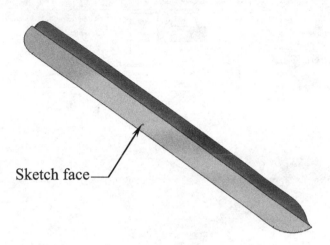

Sketch face

- Select the bottom face of the part and click **Convert Entities** [⬚].

- Add a Centerline [⬚] as shown to help position the tool when it is placed on a sheet metal part.

Convert the edges
of the Louver

Add Centerline

- **Exit** the sketch or click [⬚].

- Re-name the sketch to **Position Sketch** from the FeatureManager tree.

17. Establishing the Stop and Remove faces:

- In order for the forming tools to work properly, a set of Stopping Faces and Removing Faces will have to be established prior to saving as a forming tool.

- Change to the Sheet metal tool tab and click the **Forming Tool** command.

- For **Stopping Face**, select the **bottom face** of the part.

- For **Faces to Remove**, select the **face on the right side** of the part.

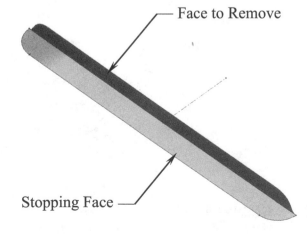

Face to Remove

Stopping Face

- Click **OK**.

18. Saving the Forming Tool:

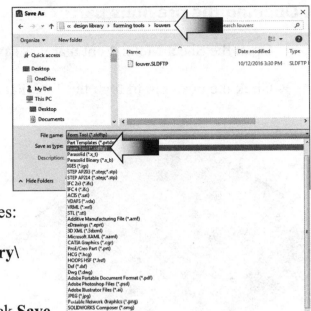

- Click: **File / Save As.**

- Enter **Long Louver** for the name of the file.

- Select **Form Tool** (*.sldftp) in the Save As Type.

- Browse to the following directories:
**Program Data\ SolidWorks\
SolidWorks 2018\ Design Library\
Forming Tools\Louvers.**

- Expand the Louver folder and click **Save**.

NOTE: *The Design Library is a Hidden Folder (Explorer / Organize / Folder and Search Options / View / Show Hidden Files, Folders and Drives).*

- After the forming tool is saved, it can be accessed through the Task pane by dragging and dropping it from the Design Library folder.

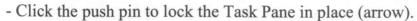

19. Opening the previous part:

- Open the Sheet Metal Vent that was saved earlier (**Alt+Tab**).

- Click the push pin to lock the Task Pane in place (arrow).

- Expand the **Design Library** folder.

- Expand the **Forming Tool** folder and double click on the **Louvers** folder to see its content.

- Hover the mouse cursor over the name Long Louver to see the preview of the Forming tool.

Forming Tools

Forming tools should be inserted from the Design Library window and applied onto parts with sheet metal parameters such as material thickness, bend allowance, fixed face, cut relief, etc...

20. Applying the form tool:

- **Drag** the Long Louver form tool from the Design Library and **drop** it approximately as indicated.

- Change the Angle to **180°** and click **Flip Tool**.

Place the Louver here

21. Positioning the form tool:

- Click the **Position** tab (arrow).

 Push / Pull

While dragging the Forming Tool from the Design Library, (still holding the mouse button) press the TAB key to reverse the direction from push to pull.

- Add the locating dimensions to the outer edges of the model to fully define this sketch.

- Use the <u>outer edges</u> of the sheet metal part when adding the dimensions.

- Click **OK**.

- The Louver is formed.

22. Adding a mounting hole:

- Select the <u>upper face</u> of the part as indicated.

- Click [⊏] to open a new sketch or select **Insert / Sketch**.

Sketch Face

- Sketch a circle [⊙] on the left side of the louver.

- Add dimensions [↖] to define the circle.

Ø.325

1.000

1.000

23. Extruding a cut:

- Click or select **Insert / Cut / Extrude**.

- End Condition: **Blind**.

- Link to Thickness: **Enabled**.

- Normal Cut: **Enabled**.

- Click **OK**.

24. Creating a Linear Pattern:

- Click or select **Insert / Pattern Mirror / Linear Pattern**.

- Select the **bottom edge** as Pattern Direction.

- Enter **.703 in**. as Spacing.

- Enter **10** as Number of Instances.

- Select the **Long Louver** as Features to Pattern.

- Click **OK**.

- The completed Linear Pattern.

25. Creating an Axis:

- An axis can be created at anytime so features can be arrayed around it. In this case, an axis in the center of the part will be made and used as the center of the next Circular Pattern.

- Click [icon] or select **Insert / Reference Geometry / Axis**.

- Click the **Two Planes** option [icon].

- Select **Front** and **Right** planes from FeatureManager tree.

- Click **OK**.

New Axis

26. Creating a Circular Pattern:

- Click or select **Insert / Pattern Mirror / Circular Pattern**.

- Select the new **Axis** for Pattern Axis.

- Enter **360 deg**. for Pattern Angle.

- Enter **4** for Number of Instances.

- For Feature to Pattern select the **Cut-Extrude1**,
  LPattern1 and the **Long Louver** from the Feature tree.

- Click **OK**.

27. Saving your work:

- Save the model with the same file name and override the previous file.

Finished Part (Folded)

Flat Pattern*

* After the formed features are created they will remain in their formed shapes even when toggled back and forth between Folded or Flattened.

Questions for Review

Sheet Metal Parts

1. The mid-point relation can be used to center a line onto the Origin.
    a. True
    b. False

2. The base flange command can also be selected from Insert / Sheet Metal / Base Flange.
    a. True
    b. False

3. When a linear model edge is selected and the sketch pencil is clicked, the system creates a plane Normal To Curve automatically.
    a. True
    b. False

4. The Miter Flange feature can create more than one flange in the same operation.
    a. True
    b. False

5. The Flat and the Folded patterns cannot be toggled until the part is completed and saved.
    a. True
    b. False

6. An existing forming tool cannot be edited or changed; forming tools are fixed by default.
    a. True
    b. False

7. When applying a form tool onto a sheet metal part, the push or pull direction can be toggled when pressing:
    a. Up arrow
    b. Tab
    c. Control

8. The Modify Sketch command can be used to rotate or translate the entire sketch.
    a. True
    b. False

9. A formed feature(s) cannot be copied or patterned.
    a. True
    b. False

	9. FALSE
8. TRUE	7. B
6. FALSE	5. FALSE
4. TRUE	3. TRUE
2. TRUE	1. TRUE

CHAPTER 14

Sheet Metal Forming Tools

Sheet Metal Forming Tools
Button with Slots

Forming tools act as dies that bend, stretch, or otherwise form sheet metal.

SOLIDWORKS includes some sample forming tools to get you started. They are stored in *Installation_Directory/Data/Design Library/FormingTools/folder_name*.

Some types of form features, such as louvers and lances, create openings on sheet metal parts. To indicate which faces of the form tool will create the openings, the system changes the color of these faces to **red,** the stopping faces to **blue** and the rest to **yellow.**

The user can only insert (drag & drop) forming tools from the **Design Library** window and apply them only to sheet metal parts. The Design Library window gives you quick access to the parts, assemblies, library features, and form tools that are used most often.

Users can create their own forming tools and apply them to sheet metal parts to create form features such as louvers, lances, flanges, and ribs.

The Design Library window has several default folders. Each folder contains a group of palette items displayed as Thumbnail Graphics.

Design Library can include:
- Parts (.sldprt)
- Assemblies (.sldasm)
- Sheet Metal Forming Tools (.sldftp)
- Library Features (.sldlfp)

In this first half of the chapter we will learn how to create and save a forming tool.

Sheet Metal Forming Tools
Button with Slots

Dimensioning Standards: **ANSI**

Units: **INCHES** – 3 Decimals

Tools Needed:

 Insert Sketch

 Boss/Base Revolve

 Convert Entities

 Dimension

 Add Geometric Relations

 Extrude Cut

 Split Line

 Fillet/Round

 Forming Tool

1. Creating the base block:

- Select the <u>Right</u> plane and Insert a new Sketch .

- Sketch a **Center-Rectangle** that centered on the origin.

- Add the width and height dimensions as shown.

(The sketch should be fully defined at this point).

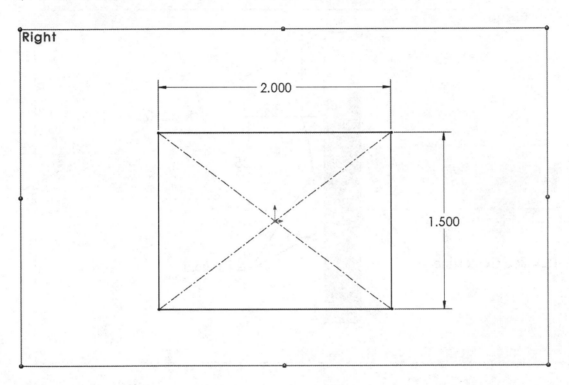

2. Extruding the base:

- Click **Extruded Boss/Base**.

- Enter the following:

- Direction 1: **Blind**.

- **Reverse** Direction.

- Depth: **.250in**.

- Click **OK**.

3. Creating the forming tool body:

- Select the <u>Front</u> plane and open a new sketch .

- Sketch the profile shown below.

- Add dimensions and relations to fully define the sketch.

4. Revolving the body:

- Click **Revolved Boss/Base** .

- Revolve **Blind** with a full **360 deg**.

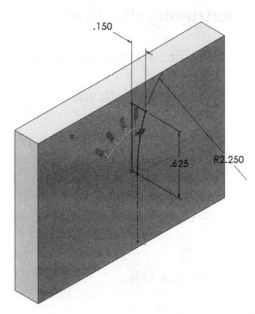

- Click **OK**.

5. Adding a fillet:

- Click **Fillet** 🔲 and enter **.080 in**. for Radius.

- Select the edge as indicated.

- Click **OK**.

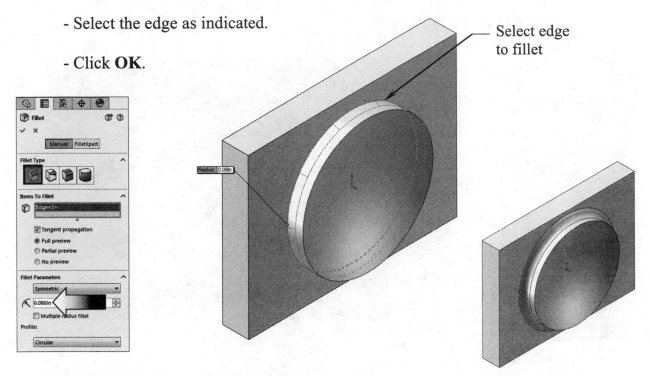

Select edge
to fillet

6. Sketching the 1st slot profile:

- We will take a look at two different methods to create the slots. For the 1st Arc-Slot, let us try out the **Offset Entities** option.

- Select the **face** as indicated and open a new sketch. Draw a Center-Point-Arc 🔲 approximately as shown.

Sketch Face

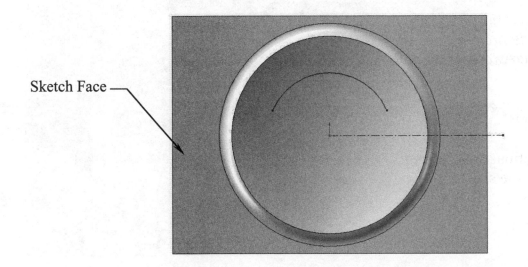

- While the arc is still highlighting, click the **Offset Entities** command.

- Enter **.0625 in**. for Offset Distance.

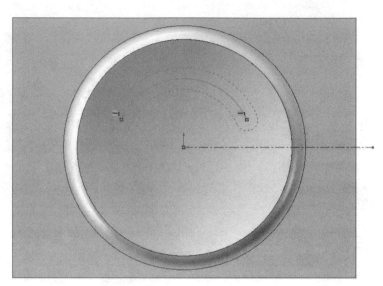

- Enable the following:

        * Add Dimensions   *** Select Chain**   *** Bi-Directional**

        * Cap End / Arcs.   *** Base Geometry** (to convert the arc to construction)

- Click **OK**.

Note:

- *We will use the Arc-Slot command to create the 2nd slot.*

- Add dimensions and the

horizontal relations shown to fully define the sketch.

135.00°

.063

.063

R.388

Horizontal

7. Creating the 1st Split Line:

- Click **Split Line** from the **CURVES** toolbar or select: **Insert / Curve / Split Line.**

- Select the <u>face</u> as indicated to split.

- Click **OK**.

Face to Split

- The selected face is split into a new, separate surface.

- This new surface can now be used as Faces-to-Remove, when the form tool is inserted onto a sheet metal part.

- The Faces-to-Remove option specifies what features/area will get a through cut.

8. Creating the 2nd slot profile:

- This time we will try another method to create the 2nd curved slot. (The Copy & Paste option also works well.)

- Select the **Face** as noted and open a new sketch.

- Select the **Center Arc Slot** under the Straight-Slot button.

- Enable the **Add Dimensions** checkbox.

- Start at the Origin and click **point 1**, move outward and click **point 2**, swing the cursor to the other side and click **point 3**, then drag down or upward to **point 4**.

- The arc-slot is completed with the Radius, Width, and Angular dimensions.

Add a horizontal centerline

Sketch face

- Change the values of the dimensions to:

  * Width: **.125 in.**
  * Radius: **R.325 in.**
  * Angle: **135 deg.**

- Create the **2nd split line** as shown in step 8.

9. Adding more fillets:

- Click **Fillet** 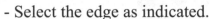 and enter **.062 in**. for radius.

- Select the edge as indicated.

- Click **OK**.

10. Inserting the Forming Tool feature:

- Select **Insert / Sheet Metal / Forming tool** from the pull-down menu.

- Select the **Stopping-Face** and the **Faces-to-Remove** as indicated.

Stopping Faces

Faces to Remove:
Creates the openings
when the form tool
is applied onto the
sheet metal parts.

- Click **OK**.

- The completed form tool.

11. Saving the Forming tool:

- Click **File / Save As / Button w-Slots**, change the Save as Type to **Form Tool** and save the part in the following directories:

C:/Program Data/SolidWorks SolidWorks 2019/Design Library /Forming Tools.

Note:

If the Design Library Folder is Hidden, go to:

Windows Explorer/ Organize/ Folder and Search Options.

*Click the **View** tab and enable the option: Show Hidden Files, Folders and Drives.*

Note:

> - _The Sheet metal forming tools can also be saved by dragging and dropping from the FeatureManager tree to a folder (i.e. Forming Tools) inside the Design Library._
>
> - _The file name, file type, and path can be selected to save the forming tool at this time._

12. Applying the new forming tool: (Optional)

> - Create a sheet metal part using the drawing below and test out your new forming tool.

Questions for Review

Sheet Metal Forming Tools

1. Forming tools can bend or stretch sheet metal parts.
   a. True
   b. False

2. Forming tools can be stored in the Design Library window using the file extension:
   a. slddrw
   b. sldftp
   c. dwg

3. Forming tools can be dragged and dropped from the Design Library window.
   a. True
   b. False

4. Forming tools can be used to form surfaces and solid parts as well.
   a. True
   b. False

5. The Red color on the face(s) of the forming tool creates openings on the sheet metal parts.
   a. True
   b. False

6. The Split Line command divides a selected face into multiple separate faces.
   a. True
   b. False

7. Only one single closed sketch can be used with the Split Line command.
   a. True
   b. False

8. The _____ key is used to reverse the direction of the forming tool (push/pull):
   a. Shift
   b. Tab
   c. Alt

7. TRUE
8. b
5. TRUE
6. TRUE
3. TRUE
4. FALSE
1. TRUE
2. B

Designing Sheet Metal Parts

Designing Sheet Metal Parts
Mounting Tray

Sheet metal components are normally used as housings or enclosures for parts or to strengthen and support other parts.

A Sheet Metal part can be created as a single part or it can also be designed in the context of an assembly that has enclosed components.

Forming tools are dies that can bend, stretch, or form sheet metal.

In SOLIDWORKS, forming tools are applied using the "Positive Half" (the raised side) to form features.

When inserting a forming tool, its direction can be reversed using the TAB key (Push or Pull).

The Sheet Metal part can be flattened either by using the Unfold or Flattened button, and drawings can be made to show views of the bent or flattened part. Bend lines are also visible in the drawing views.

By default, only the Bend-Lines are visible at all time, but not the Bend-Regions. To show the Bend Regions, right click on the Flat Pattern1 icon at the bottom of the FeatureManager tree, then select Edit Feature and <u>clear</u> the <u>Merge Faces</u> check box.

In this second half of the chapter, besides learning how to create a sheet metal part, we will also learn how to apply the form tool that was created earlier in the first half of the lesson.

Designing Sheet Metal Parts
Mounting Tray

Dimensioning Standards: **ANSI**
Units: **INCHES** – 3 Decimals

Tools Needed:

Insert Sketch	Base Flange	Edge Flange
Unfold	Fold	Extruded Cut
Linear Pattern	Flattened	Break Corner

1. Starting with the base sketch:

- Select the <u>Front</u> plane from FeatureManager Tree.

- Click or select **Insert / Sketch**.

- Sketch the profile below and add dimensions to fully define the sketch.

Front Plane

2.000

1.000

Base Flange

A Base Flange is the first extruded feature in a sheet metal part.
Sheet metal parameters are added automatically.

2. Extruding the Base Flange:

- Click on the Sheet Metal toolbar, or select:

   Insert / Sheet Metal / Base Flange.

- End Condition: **Blind, Material Outside.**

- Extrude Depth: **5.000.**

- Thickness: **18 Gauge.**

- Override Radius: **Enabled.**

- Bend Radius: **.010.**

- Click **OK.**

3. Creating an Edge Flange:

- Select the **outer edge** as shown and click on the Sheet Metal toolbar or select:
Insert / Sheet Metal / Edge Flange.

Select the
outer edge

> ⚡ **Edge Flange**
>
> The Edge Flange command adds a flange to the selected linear edge and shares the same material thickness as the sheet metal part.

- Position the flange towards the left side and set the following:

* Use Default Radius: **Enabled**.

* Flange Direction: **Blind**.

* Bend Angle: **90deg**.

* Flange Depth: **.985**.

* Use **Inner Virtual Sharp** (arrow).

* Flange Position: **Material Outside**.

- Click **OK**.

4. Adding cut features:

- Select the <u>face</u> as shown and insert a new sketch 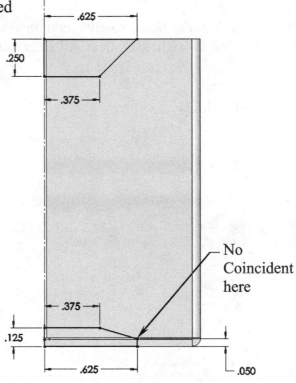.

- Sketch the profile and add dimensions.

- The horizontal dimensions are measured from the Centerline.

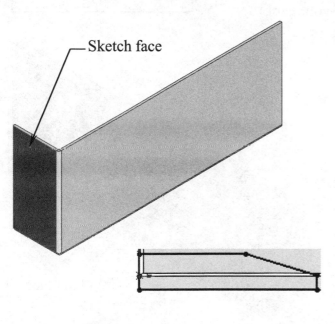

Sketch face

No Coincident here

.625
.250
.375
.375
.125
.625
.050

5. Extruding a cut:

- Click or select **Insert / Cut / Extrude**.

- End Condition: **Blind**.

- Extrude Depth: **.165 in.**

- Click **OK**.

6. Using the Unfold command:

- Click on the Sheet metal toolbar or select:
Insert / Sheet Metal / Unfold.

- Select the **right face** as Fixed face.

- Select the **bend radius** as indicated for Bends to
Unfold. (Click Collect All Bends if you want to
flatten the entire part.)

- Click **OK**.

When adding cuts across
a bend, the Unfold
command flattens one or
more bend(s) in a sheet
metal part.

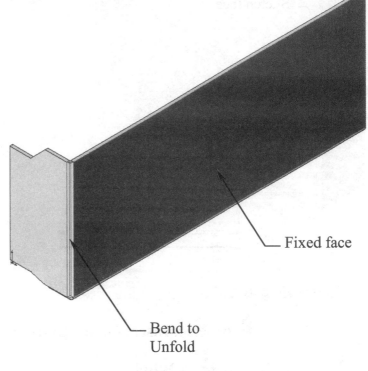

Fixed face

Bend to
Unfold

Unfold

7. Creating a Rectangular Window:

- Select the <u>face</u> as indicated and insert a new sketch .

- Sketch a **Corner rectangle** ⬜ as shown below.

- Add dimensions ◇ and Sketch Fillets ⌐.

Symmetric

Sketch face

R.032

.875

.625

.500

8. Extruding a Cut:

- Click 🔲 or select **Insert / Cut / Extrude**.

- End Condition: **Blind**.

- Link to Thickness: **Enabled** (default).

- Normal Cut: **Enabled**.

- Click **OK**.

R.032

.875

.625

.500

💡 **Link To Thickness**

- The Link-to-Thickness option is only available in sheet metal parts.

- When this option is enabled, the extruded depth is automatically linked to the part thickness.

---

9. Using the Fold command:

- Click on the Sheet metal toolbar or select:

    Insert / Sheet Metal / Fold.

- Select the **right face** as Fixed face.

- Select the **bend radius** as indicated for Bends to Fold. (If Collect-All-Bends was selected last time, click it again this time.)

- Click **OK**.

UnFold

Fixed face

Bend to Fold

Hide the Sketches

Folded

10. Unfolding multiple bends:

- Click 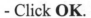 or select **Insert / Sheet Metal / Unfold**.

- Select the <u>bottom face</u> as Fixed face.

- Select the **faces** of the 2 bends as shown for Bends-To-Unfold.

- Click **OK**.

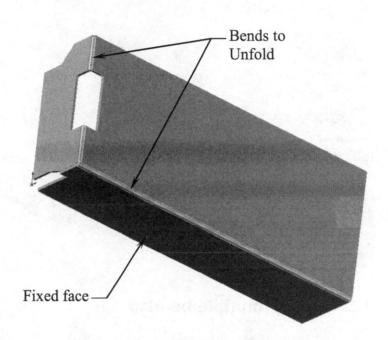

Bends to Unfold

Fixed face

11. Adding more Cuts:

- Select the **upper face** as noted and insert a new sketch.

- Sketch **2 Circles** and add dimensions for size and position.

- Add an **Equal** and a **Vertical** relation between the circles.

Sketch face

Ø.500

1.000

1.125

2.500

- Click or select **Insert / Cut / Extrude**.

- End Condition: **Blind**.

- Link To Thickness: **Enabled**.

- Click **OK**.

12. Folding multiple bends:

- Click on the Sheet Metal toolbar or select: **Insert / Sheet Metal / Fold**.

- The Fixed face is still selected by default.

- Under Bends to Fold, click **Collect-All-Bends**.

- Click **OK**.

Unfold

Folded

13. Inserting a Sheet Metal Forming Tool:

- Click the **Design Library** icon and click the push pin to lock it.

> ### Forming Tools
>
> Forming tools are dies that can bend, stretch, or form Sheet Metal and they must be inserted from the Design Library window.

- Expand the **Design Library** folder.

- Click on the **Forming Tools** folder.

- Click on the **Lances** folder.

- Locate the **Bridge Lance** form tool.

- Hover the mouse cursor over the Bridge Lance icon to see its preview graphics.

- SOLIDWORKS includes some sample forming tools to help get you started. These form tools can be customized and used in different sheet metal parts.

- Drag the Bridge Lance* from the Task Pane and drop it on the sheet metal part approximately as shown.

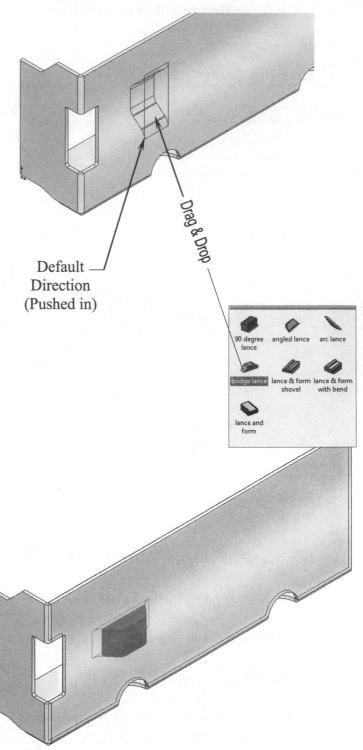

- By default, this form tool is inserted inwards (pushed in) and orientated vertically.

Default Direction (Pushed in)

Drag & Drop

- To correctly position the form tool, change the Rotation Angle to **180°** and click the **Flip Tool** button (arrows).

If the Bridge Lance fails to form, double click on its icon to open the actual part, and re-save it in the same location, but using the new Forming Tool extension (.sldftp)

14. Locating the Bridge Lance:

- Click the **Position** tab (arrow).

- Add the dimensions shown below to correctly position the formed feature.

- Dimensions <u>from the centerlines</u> to the outer-most edges of the part.

- Click [Finish] .

- Un-pin the Design Library tree to put it away temporarily.

15. Creating the Linear Pattern of the Bridge Lance:

- Click or select **Insert / Pattern Mirror / Linear Pattern**.

- Select the top **horizontal edge** of the part as Pattern Direction.

- Enter **1.500** in. for Instance Spacing.

- Enter **3** for Number of Instances.

- Select the **Bridge-Lance** as Features to pattern.

- Click **OK**.

Direction arrow

16. Mirroring the body:

- Rotate the part and select the face as indicated (for mirror face).

Mirror face

- Select **Insert / Pattern Mirror / Mirror**.

- Expand the **Bodies to Mirror** section and select the body as shown.

- Click **OK**.

Body to mirror

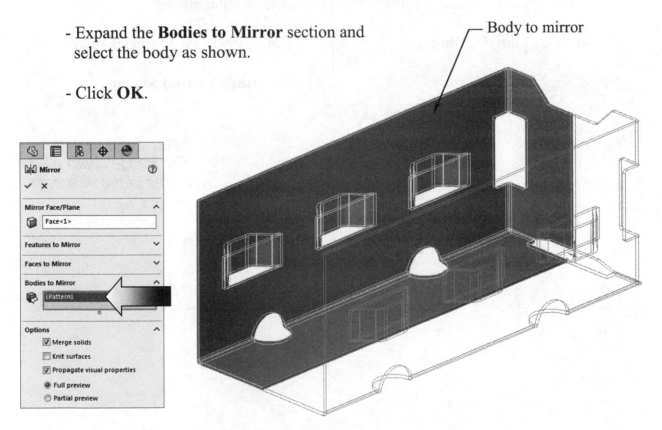

17. Adding the rear Edge Flange:

- Select the <u>edge</u> as indicated.

- Click 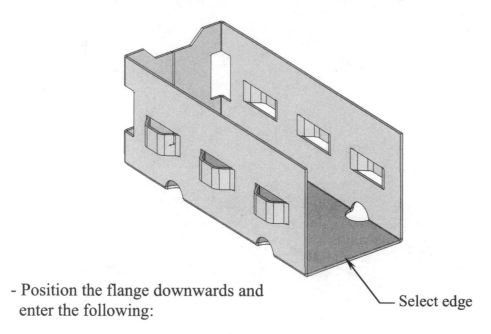 or select **Insert / Sheet Metal / Edge Flange**.

— Select edge

- Position the flange downwards and enter the following:

- Use Default Radius: **Enabled**.

- Flange Length: **Blind**.

- Bend Angle: **90deg**.

- Flange Position: **Material Inside**.

- Use **Inner Virtual Sharp**.

Note:
The Flange depth will be edited in the next step.

18. Resizing the Edge Flange:

- Click the **Edit Flange Profile** button (arrow).

- The 2D sketch of the flange is activated so that its shape and size can be altered.

- **Drag** the 2 outer-most vertical lines inward (pictured).

- Sketch **2 Circles** and add the Dimensions as shown.

- Click to **exit** the sketch

or click **Rebuilt**.

19. Adding Chamfers:

- Click or select **Insert / Sheet Metal / Break-Corner**.

- Break Type: **Chamfer**.

- Enter **.060** for chamfer depth.

- Select the **4 Edges** as shown.

- Click **OK**.

Select 4 Edges

Front Isometric

Back Isometric
See note below

Rotate Option 1: Shift + Up or Down Arrow <u>twice</u> to rotate 90° each time.
Rotate Option 2: Set the View rotation to 15 degrees – Press the **Right** arrow 12 times and the **Down** arrow 4 times.
OR: use the new option **View Selector** (Space Bar) and click one of the faces (or projected faces) on the cube to rotate the model to that orientation.

20. Switching to the Flat Pattern:

- Click [icon] or select **Insert / Sheet Metal / Flattened***.

- Verify that the part is flattened properly and there are no rebuild errors.

Fixed face

* *To change the Fixed face of the part, edit the* ***Flat-Pattern1*** *feature and select the face as noted and press OK.*

21. Saving your work:

- Select **File / Save As / Mounting Tray / Save**.

Questions for Review

Sheet Metal Forming Tools

1. A Sheet Metal part can be created as a single part or in context of an assembly with enclosed components.
    a. True
    b. False

2. A Base Flange is the first extruded feature in a sheet metal part. Sheet metal parameters are added automatically.
    a. True
    b. False

3. A sheet metal part designed in SOLIDWORKS can have multiple wall thickness.
    a. True
    b. False

4. The Edge Flange command adds a flange to the selected linear edge and shares the same material thickness of the sheet metal part.
    a. True
    b. False

5. Only one bend can be flattened at a time using the Unfold command.
    a. True
    b. False

6. Forming tools have to be inserted from the Feature Palette window.
    a. True
    b. False

7. To reverse the direction of the forming tool while being dragged from the Feature Palette window, press:
    a. Tab
    b. Control
    c. Shift

8. After the features are created by the forming tools, their sketches can only be moved or re-positioned, and their dimension values cannot be changed.
    a. True
    b. False

1. TRUE	2. TRUE
3. FALSE	4. TRUE
5. FALSE	6. TRUE
7. A	8. TRUE

CHAPTER 15

Sheet Metal Conversions

Sheet Metal Conversions
From IGES to SOLIDWORKS

Parts created from other CAD systems (using IGES or **I**nitial **G**raphics **E**xchange **S**pecification) can be imported and converted into SOLIDWORKS Sheet Metal.

When importing other CAD formats into SOLIDWORKS, the software recognizes them as follows:

* If there are blank surfaces, they are imported and added to the Feature Manager design Tree as surface features.
* If the attempt to knit the surfaces into a solid succeeds, the solid appears as the base feature (named **Imported1**) in a new part file.
* If the surfaces represent multiple closed volumes, then one part is generated for each closed volume.
* If the attempt to knit the surfaces fails, the surfaces are grouped into one or more surface features (named **Surface-Imported1...**) in a new part file.
* If you import a **.dxf** or **.dwg** file, the **DXF/DWG import wizard** appears to guide you through the import process.

The imported parts must be of uniform thickness to fold and unfold properly.

After the part is opened in SOLIDWORKS, there are several methods to convert it to a sheet metal part, but the sheet metal parameters such as Rip, Fixed face or edge, Bend radius, etc., must be added before the Flat Pattern can be created.

The converted part appears on the Feature Manager Design tree, it contains the features Sheet Metal1, Flatten Bend1, and Process Bend1.

The sheet metal part can now be Flattened and Folded by toggling the Suppression state of the Process Bends.

Sheet Metal Conversions
From IGES to SOLIDWORKS Flat Pattern

Dimensioning Standards: **ANSI**

Units: **INCHES** – 3 Decimals

Tools Needed:

Convert to Sheet Metal	Insert Bend	Flat Pattern
Sheet Metal Gusset	Flatten Bend	Process Bends

1. Opening an IGES document:

- Go to **File / Open**. Change Files of Type to **IGES**.

- Browse to the Training Files folder and open: **Sheet Metal Conversion**.

- Click **No** to skip the Import-Diagnostics option.

- Click **No** to skip the Feature-Recognition option.

- The part is imported into **SOLIDWORKS** as solid body with no feature history.

- Change the Display Style to: **Shaded-With Edges**.

2. Creating the Rips:

- Click **Rip** on the sheet met
toolbar or select **Insert / Sheet Metal / Rip**.

- Select the **inner edge** as shown.

- The 2 arrows indicate that the Rip command
is going to cut both walls.

Select the
1st edge.

RIP

The RIP command
creates a tear between the
2 walls. A Gap is entered
and the rip direction can
be toggled to achieve the
Butt, Over-lapped, or
Under-lapped results.

- Click on the direction arrow as noted, to keep the correct
Sides from being ripped.

- Use the **Default Gap** (.010).

Either click on
Change Direction
-OR- click on the
arrow itself.

Default

Direction 1
(Over lapped)

Direction 2
(Under lapped)

- Select a total of **4 edges** (2 on each side) as indicated.

- Click **OK**.

Arrows indicate
the trimmed side

Select
4 edges

3. Inserting the Sheet Metal Parameters:

- Click **Insert Bends** command or select **Insert / Sheet Metal / Bends**.

- Select the inside **face** to use as the Fixed Face.

- Enter **.015 in.** for inside Bend Radius.

- Click **OK**.

- The system reports some Auto Relief Cuts were made for some bends.

- Click OK .

Fixed Face

4. Adding Fillets:

- Click **Break Corner / Corner Trim** command and select the **Fillet** option.

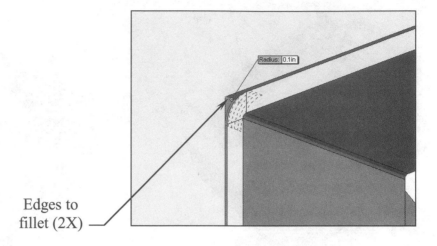

Edges to fillet (2X) —

- Enter **.100 in**. for Radius.

- Select the **2 edges** as noted.

- Click **OK**.

- Rotate the model to verify the resulted fillets.

5. Switching to the Flat pattern:

- To examine the part in the flattened view; click the **Flatten** command on the Sheet Metal toolbar (arrow).

6. Saving your work:

- Click **File / Save As**.

- Enter **Sheet Metal Conversion** for the file name.

- Click **Save**.

Questions for Review

Sheet Metal Conversions

1. An IGES file can be imported into SOLIDWORKS and converted into a sheet metal part.
    a. True
    b. False

2. DXF and DWG are imported into SOLIDWORKS as 2D Sketches, using the DXF/DWG Import-Wizard.
    a. True
    b. False

3. After being imported into SOLIDWORKS, the IGES file can be flattened instantly.
    a. True
    b. False

4. The imported parts must be of uniform thickness to fold and unfold properly.
    a. True
    b. False

5. The Rip feature removes 1 material thickness based on the side of the direction arrow that you select.
    a. True
    b. False

6. When applying the sheet metal parameters you do not have to specify a fixed face.
    a. True
    b. False

7. The width and depth of the relief cuts are fixed and cannot be changed.
    a. True
    b. False

8. The Folded and the Flat pattern can be toggled by moving the Rollback Line up or down.
    a. True
    b. False

7. FALSE   8. TRUE
5. TRUE   6. FALSE
3. FALSE   4. TRUE
1. TRUE   2. TRUE

Sheet Metal Gussets

Sheet Metal gussets can now be created in SOLIDWORKS with specific indents that go across bends. This exercise will guide us through the creation of a gusset in a sheet metal part.

1. Opening a sheet metal part document:

- Click **File / Open**.

- Browse to the Training files folder and open a part document named: **Sheet-Metal Gusset**.

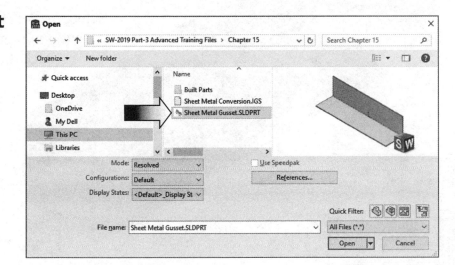

2. Creating a gusset:

- Click the **Sheet Metal Gusset** command from the Sheet Metal tool tab (arrow).

- Under the **Position** section, select the **2 Faces** as indicated in the balloon number 1 for Supporting Faces.

- Select the **horizontal edge** in the middle of the 2 supporting faces as noted in the balloon number-2 to be used as a reference to align the gusset.

- Select the **Vertex** as noted in the balloon number 3 to be used for dimensioning.

Reference edge ②

Reference point ③

2 supporting faces ①

2.000 ④

Sheet Metal Gusset

Position

Face<1>
Face<2> ①

Edge<1> ②

Point<1> ③

☑ Offset

2.000in ④

- Move down to the **Profile** section and click the **Profile Dimensions** option (arrow).

- Enter **.500 in**. for Profile Length.

- Enter **.500 in**. for Profile-Height.

- Click the **Rounded Gusset** button to create a gusset with a rounded edge.

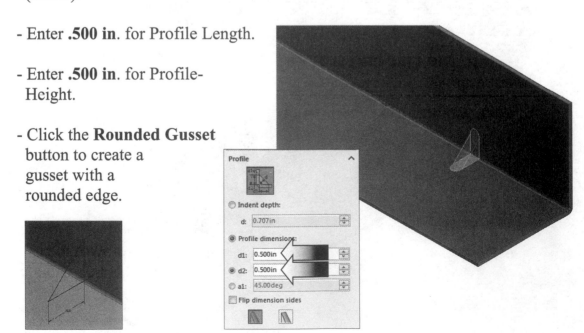

Profile

○ Indent depth:
   d: 0.707in

● Profile dimensions:
   d1: 0.500in
   ● d2: 0.500in
   ○ a1: 45.00deg

☐ Flip dimension sides

- Move down to the Dimensions section and enter the following:

    * Indent Width Dimension: **.1875 in**.

    * Indent Thickness Dimension: **.060 in**.

    * Inner Corner Fillet: **.030 in**.

    * Outer Corner Fillet: **.090 in**.

- Enable the **Full Preview** option in the Preview section, if needed.

- Click **OK** to create the gusset.

3. Viewing the resulted gusset:

- Zoom in on the new gusset to see its details. Also rotate the view to see the indent from the back side.

- Click on the feature itself to see its dimensions.

- The gusset has a built-in sketch that can be edited to change to a custom profile if needed.

4. Mirroring the gusset:

- Change to the Features tool tab and click the **Mirror** command.

- Select the **Right** plane from the Feature tree for Mirror-Plane.

- Select the **Gusset1** also from the Feature tree for Features-to-Mirror.

- Click **OK**.

5. Saving your work:

- Click **File / Save As**.

- Enter: **Sheet Metal Gusset** for the name of the file.

- Click **Save**.
Click Yes to replace the old file with the new when prompted.

Flat Pattern Stent

Flat Pattern Stent
Flat Pattern Approach

Using the built-in Sheet Metal features in SOLIDWORKS you can flatten or roll solid models such as wire mesh screens, grill meshes, or stent patterns.

When designing a sheet metal part, the material setback is something we must keep in mind: The Bend allowance and bend deduction calculations are methods you can choose to determine the flat length of sheet stock to give the desired dimension of the bent part. This lesson uses the default settings of the K-Factor to calculate the bend allowance (BA=P(R + KT) A/180).

K-Factor is a ratio that represents the location of the neutral sheet with respect to the thickness of the sheet metal part. When you select K-Factor as the bend allowance, you can specify a K-Factor bend table. The SOLIDWORKS application also comes with a K-Factor bend table in Microsoft Excel format.

There are several known methods for creating these types of patterns; this lesson will walk you through the use of rolling and unrolling a cylinder and its pattern using the Sheet Metal functions in SOLIDWORKS.

Flat Pattern Stent
Flat Pattern Approach

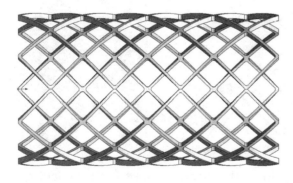

Dimensioning Standards: **ANSI**
Units: **INCHES** – 3 Decimals

Tools Needed:

	Revolve Boss-Base		Flatten		Unfold
	Fold		Extruded Cut		Fillet / Round

1. Starting with a new part document:

- Click **File / New / Part**, set the units to **Inches**, **3 decimal** places.

- Select the <u>Front</u> plane and open a new sketch.

- Create the sketch and add the dimensions / relation shown below.
  (The dimensions are scaled up for ease of modeling purposes.)

Front Plane

.020

1.040

Horizontal

Ø.600

- Click **Revolve / Boss-Base** .

- The revolve centerline is selected automatically.

- Use the default **Blind** type.

- Enter **359.9deg** for angle.

- Click **OK**.

  (The gap is needed to flatten the part later on.)

.020

1.040

Ø.600

2. Converting to Sheet Metal :

- Click the **Insert-Bends** command on the Sheet Metal toolbar.

- Select the left <u>edge</u> as noted for Fixed-Edge/Face.

- Enter **0** for bend radius.

- Use the default K-Factor and Auto Relief settings.

Select the Fixed edge

- Click **OK**.

- The solid model is converted to a Sheet Metal part. Press the Flatten command on the Sheet Metal toolbar to see its flat pattern.

- Click the flat pattern command again to roll the part back to its default shape.

3. Unfolding the part:

- Select the **Unfold** command from the Sheet Metal toolbar.

- Select the same <u>edge</u> to keep as the Fixed Edge/Face.

- Click the **Collect-All-Bends** button (arrow).

- Click **OK**.

- The part is flattened but this time new features can be added and they will roll back when the part is folded.

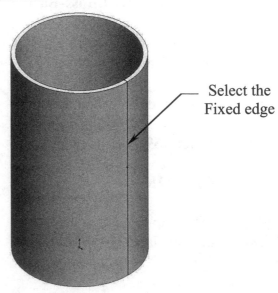

Select the Fixed edge

4. Adding the sketch pattern:

- Select the <u>face</u> as noted and open a new sketch.

- Sketch a couple of squares (notice the upper square is slightly larger than the lower one by .006").

- Add the dimensions and relations shown to fully define the sketch.

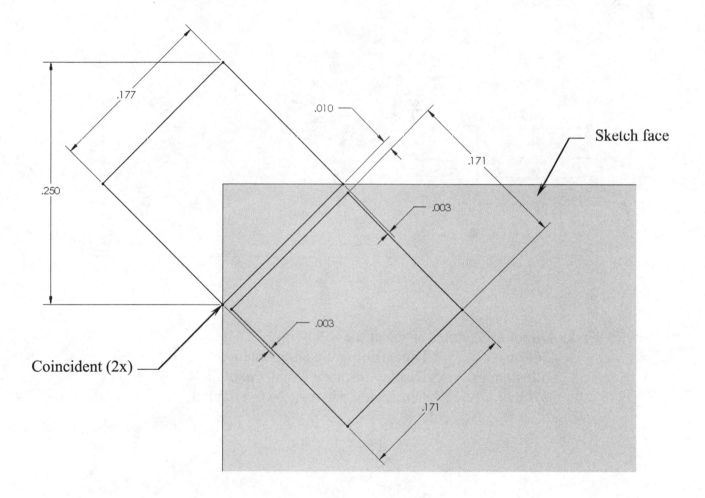

- Additional relations such as Parallel, Equal, or Perpendicular can also be used to help eliminate some dimensions

- We are going to use the Linear Sketch Pattern command to repeat the two squares several times, so there are a few things to keep in mind:

    * Pre-select the entities to pattern.   *Auto add spacing dimensions.
    * Pattern along 2 directions, use angle (0deg and 270deg for directions).

- Select both squares and click the **Linear-Sketch-Pattern** command from the Sketch toolbar, or select it from the menus: **Tools / Sketch Tools**.

- Under **Direction 1**, enter / select the following:
  * **.260in**   * Dimension X Spacing enabled.
  * **8** Instances   * Display Instance Count enabled.
  * **0deg**

- Under Direction 2, enter or select the following:
  * **.260in**   * Dimension Y Spacing enabled.
  * **5** Instances   * Display Instance Count enabled.
  * **270deg**   * Dimension between Axes enabled.

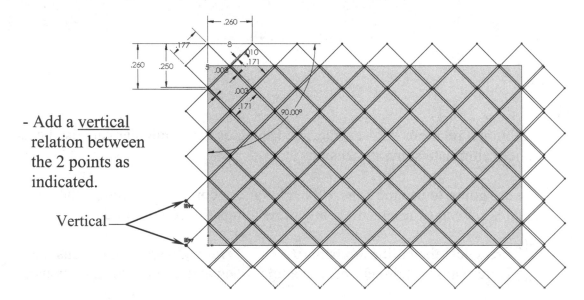

- Add a <u>vertical</u> relation between the 2 points as indicated.

Vertical

5. Creating a cut with Link to Thickness:

- Select the **Extruded Cut** command from the Sheet Metal toolbar.

- Us the default **Blind** option and enable the **Link To Thickness** checkbox.

- Click **OK**.

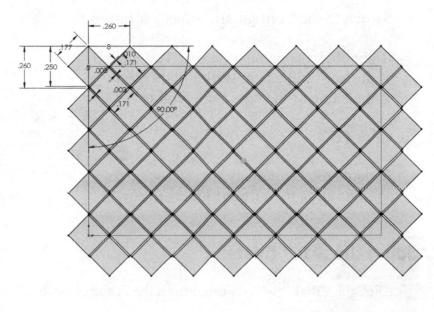

6. Folding the part:

- Click the **Fold** command from the Sheet Metal toolbar.

Select the fixed edge

- Select the small <u>vertical edge</u> as as noted for Fixed Face/Edge.

- Click the **Collect-All-Bends** button.

- Click **OK**.

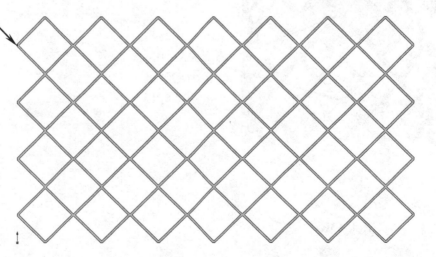

7. Creating a new configuration:

- Before adding the fillets to all the intersections, we will need to create a configuration so that the fillets can be added and captured in a separate configuration.

- Switch to the ConfigurationManager (arrow).

- Right click the name of the part (on the top of the tree) and enter: **With Fillets** (arrow) as the name of the new configuration.

- Click **OK**.

- Rename the Default configuration to **Without Fillets** (arrow).

8. Adding the .010" fillets:

- Click the **Fillet** 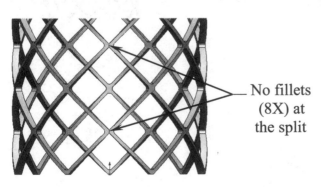 command from the Features toolbar.

- Use the default Constant radius option.

- Enter **.010in** for radius.

- Select all edges <u>except</u> for the ones at the two ends as indicated.

No fillets (8X) at the split

- Click **OK**.

9. Switching to Flatten mode:

- Click **Flatten** on
the Sheet Metal toolbar.

- The part is flattened and
the cut pattern is unrolled
with it.

- Use the Flatten command to flatten the sheet metal
part to check its dimensions, get a print out from it, or
export it as a DXF or DWG
to use in manufacturing.

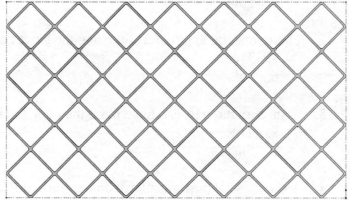

- Use the Fold and Unfold
commands to flatten the
part and add new features,
so that these features can
roll or unroll with the part.

10. Switching configuration:

- Switch back to the **Without Fillets** configuration by
double clicking on its name.

- Click the **Flatten** command again to verify the pattern.

- At this point, the pattern can be exported or a print can
be made from it for inspection or documentation
purposes.

11. Saving your work:

- Click **File / Save As**.

- Enter **Flat Pattern Stent** for the name
of the file.

- Click **Save**.

Questions for Review

Flat Pattern Stent

1. Using SOLIDWORKS, a cylinder or a cone can be unrolled into a Sheet Metal flat pattern.
   a. True
   b. False

2. There must be a gap or a slit, along the length of the cylinder for it to flatten.
   a. True
   b. False

3. A sheet metal part can have more than one thickness.
   a. True
   b. False

4. A sheet metal part must have one uniform thickness.
   a. True
   b. False

5. The Link to Thickness option links the depth-of-cut to the thickness of the part.
   a. True
   b. False

6. The K-Factor value is locked to .5; this ratio cannot be changed.
   a. True
   b. False

7. Use the Flatten command to flatten the part and add new features.
   a. True
   b. False

8. Use the Unfold command to flatten the part and add new features.
   a. True
   b. False

7. FALSE   8. TRUE
5. TRUE   6. FALSE
3. FALSE   4. TRUE
1. TRUE   2. TRUE

Exercise: Stent Example - Sheet Metal Approach

There are many known methods for creating the shapes of stents. This exercise will show the one that uses the combination of Patterns, Ribs and Combine Common options to create the model shown above. (The dimensions in the model are scaled up for visual purposes.)

1. Creating the main sketch:

- Select the <u>Front</u> plane and open a new Sketch.

- Sketch a **line** centered on the origin, and **two center-lines** as shown.

- Add the dimensions to fully define the sketch.

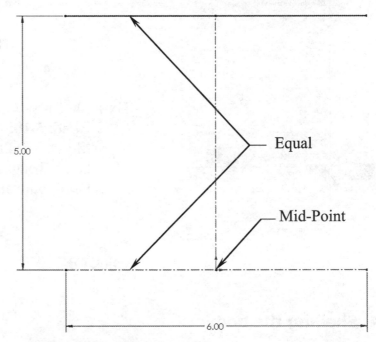

Equal

Mid-Point

5.00

6.00

2. Revolving the main body:

- From the Features toolbar, click **Revolve Boss-Base**.

- Set Direction 1 to: **Mid-Plane**.

- Set Revolve Angle to: **90deg**.

- Set the Thickness under Thin Feature to: **.040in**.

- Click **OK**.

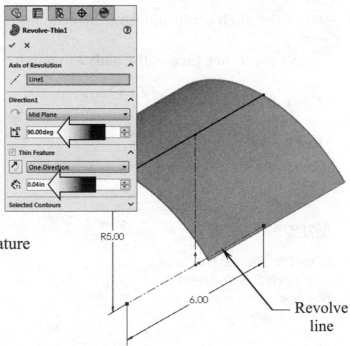

R5.00

6.00

Revolve line

3. Creating the 2ⁿᵈ body:

- Select the <u>Top</u> plane and open a new sketch.

- Sketch a **Center Rectangle** and add the dimension shown.

- **Extrude Boss-Base**:
  Blind: **5.10in**
  Merge Result: **Cleared** (arrow).
  Draft: **1deg** (arrow).
  Draft Outward **Enabled**.

- Click **OK**.

4. Shelling the body:

- Click the **Shell** command from the Features toolbar.

- Select the **upper face** of the **body2** to remove.

- Enter **.040in.** for thickness.

- Click **OK**.

<u>*Note:*</u>

Only the Body2 is shelled. The Body1 is set below the top surface by.100".

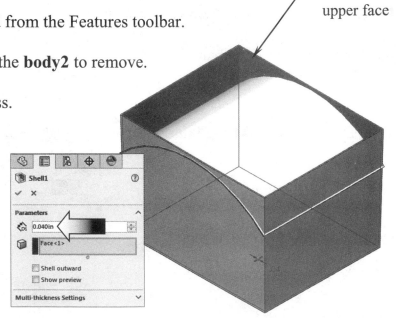

Remove the upper face

5. Creating an offset plane:

- Select the <u>uppermost face</u> of the part and create an offset plane at **.040in**.

- Click the **Flip** checkbox to place the new plane **below** the surface.

- Click **OK**.

6. Creating a Rib feature:

- Open a new sketch on the <u>new plane</u>.

- Add a line across the walls, at the upper left corner of the part.

- Sketch a centerline that is coincident to the 2 diagonal corners.

- Add the dimension and relations as noted.

- Click the **Rib** command from the Features toolbar.

- Set the thickness to **Mid Plane**.

- Set the wall to: **.040in**.

- Click the **Normal To Sketch** button (arrow).

- Under the Selected Body section, click the **Shelled** body (arrow).

- Click **OK**.

Offset the uppermost face.

.040

Perpendicular

.20

7. Patterning the Rib:

- Click **Linear Pattern**.

- Click the **front face** of the rib to see its dimensions.

- For Pattern Direction: select the **.200in** dimension.

- For Spacing, enter: **.502**in.

- For Number of Instances, enter: **18**.

Select this dimension for direction

Click the face of the Rib1

- For Features to Pattern, select the **Rib1** either from the graphics area or from the tree.

- Expand the Feature-Scope section, <u>clear</u> the Auto-Select check box, and click the <u>face of the Rib</u> as noted.

- Enable the **Vary Sketch** checkbox.

- Click **OK**.

- The rib feature is repeated 18 times along the direction that was specified by the .200in dimension.

- This is the 1st set of the ribs. We'll repeat steps 6 and 7 again to create similar ribs on the opposite side.

8. Creating another Rib:

- Select the <u>upper face</u> of the part and open a new sketch.

- Sketch a line across the walls as shown.

- Add a centerline that is coincident to the 2 diagonal corners.

- Add the dimension and relations as indicated.

- Create another rib using the <u>same settings</u> as the first one.

- Click **OK**.

9. Patterning the 2nd set of the ribs:

- Click **Linear Pattern**.

- Click the front face of the rib to see its dimensions.

- For Pattern Direction: <u>double click</u> the **.20**in. dimension.

- For Spacing, enter: **.502in**.

- For Number of Instances, enter: **18**.

- For Features to Pattern, select the **Rib1**.

- <u>Clear</u> the Auto-Select checkbox in the Feature-Scope section and click the <u>face of the Rib</u>. Also enable **Vary Sketch**.

- Click **OK**. The Rib is repeated 18 times.

Sketch Face

Perpendicular

10. Using Combine Common:

- Select **Insert / Features / Combine**.

- Click the **Common** option.

- Select **all bodies** either from the graphics area or from the Feature tree.

- Click the **Show Preview** button.

- Click **OK**.

- The Combine-Common removes all material except that which overlaps.

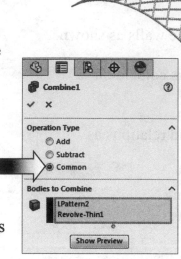

11. Saving the part:

- Save the part as **Stent Sample.sldprt**

12. Making an assembly from the part:

- The first one-quarter of the part is finished. We are going to place it in an assembly document and create 3 more instances to make the complete part.

- Click **File / Make Assembly From Part**.

- Select the **Assembly Template** and click **OK**.

- Place the component on the Origin as noted.

13. Creating a Circular Component pattern:

- Enable the <u>Temporary Axis</u> under the
  View / **Hide/Show** pull-down menus.

- On the Assembly tool tab, select:
  Circular Component Pattern (arrow).

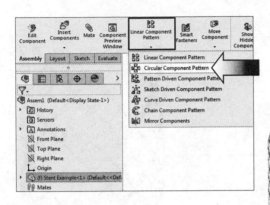

Place the part
on the Origin

- For Pattern Axis, click
  the **Temporary Axis**.

- Enable the **Equal Spacing**
  checkbox.

- Enter **4** for number of instances.

- Under Components to Pattern, select the component
  in the graphics area.

- Click **OK**.

14. Saving the assembly:

- Click **File / Save As**.

- Enter **Stent Sample Assembly.sldasm** for the
  name of the file.

- Click **Save**.

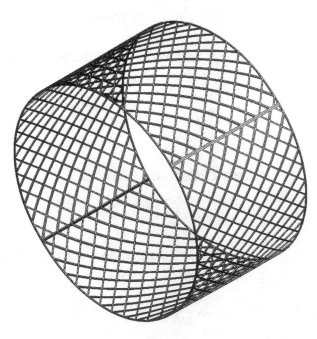

CHAPTER 16

Working with Sheet Metal STEP Files

Working with Sheet Metal STEP Files

STEP file extension is short for: **ST**andard for the **E**xchange of **P**roduct data.

The STEP translator supports import and export of body, face, and curve colors from STEP AP214 files.

The STEP AP203 standard does not have any color implementation. The STEP translator imports STEP files as SOLIDWORKS part or assembly documents.

The STEP translator exports SOLIDWORKS part or assembly documents to STEP files. You can select to export individual parts or subassemblies from an assembly tree, limiting export to only those parts or subassemblies.

If you select a subassembly, all of its components are automatically selected. If you select a component, its ascendants are partially selected, preserving the assembly structure.

This lesson discusses one of the methods to convert an Assembly STEP file into SOLIDWORKS Sheet Metal parts.

After the components are converted, some of the Assembly Features such as the Hole Series and Hole Wizards are used to add the new holes in the assembly mode; then the Fasteners are inserted automatically using the Smart Fasteners feature (requires SOLIDWORKS Toolbox).

Working with Sheet Metal STEP Files

Dimensioning Standards: **ANSI**

Units: **INCHES** – 3 Decimals

Tools Needed:

 Dimension   Insert Bends   Flat Pattern

 Hole Series   Hole Wizard   Smart Fasteners

Sheet Metal STEP Files and Smart Fasteners

1. Opening an Assembly Step File:

- Go to **File / Open**. Browse to the Training Files folder, change the Files of Type to **STEP** and open the STEP document named: **SM-Assembly.step**.

- The part files from the STEP document will appear as SOLIDWORKS documents on the FeatureManager tree without any model history.

- Click **OK** when the message Improve Free Curves appears.

- There are 4 components in this assembly and they have not yet been constrained.

- The Housing will be used as the Fixed Component and the 3 Card Guides will be left un-constrained for the purpose of this exercise.

- Change the Shading option to: **Shaded with Edges** (arrow).

Shaded with Edges

2. Mating the components:

- In order to mate the Housing component to the assembly's Origin, we'll need to align the 3 Front, Top, and Right planes.

Coincident 2 Front Planes

- Click the **Mate** command from the Assembly toolbar.

- Select the **Front** plane of the Housing and the **Front** plane of the Assembly.

- Select **Coincident** From the list.

- Click **OK**.

- Add the same Coincident mate to the **Top** plane of the Housing and the **Top** plane of the Assembly.

Coincident 2 Top Planes

- Repeat the Coincident mate for the **Right** plane of the Housing and the **Right** plane of the Assembly.

Coincident 2 Right Planes

- The Housing component should be **Fixed** (f) at this time.

3. Adding other Mates:

- Other mates can be added to constrain the other components, but later on they'll need to be suppressed so that the final sheet metal components can be flattened properly.

- In this exercise we will leave the components un-constrained to help focus on other areas.

4. Examining the components:

- The imported components have **Sharp corners** all around, which is not suitable for the Sheet Metal processes.

Sharp Corner(s)

- Sheet Metal parameters* must be added to fully convert the imported part into a SOLIDWORKS Sheet Metal part.

5. Adding the Sheet Metal tool tab:

- In an assembly, if the Sheet Metal tool tab is not visible on the CommandManager, do the following to add it:

- Right click on the Assembly tool tab and select the **Customize Command Manager** option (arrow).

- Click the **New Tab** and pick: **Sheet metal** (arrow), and click **OK**.

- Click the new **Sheet Metal** tool tab.

Sheet Metal Tab

6. Inserting Sheet Metal parameters*:

- Select the component **Housing** and click the **Edit Component** command.

- From the Sheet Metal tool tab click the **Insert Bends** command.

- Select the **Fixed face** as noted.

- Enter **.010in** for Bend Radius.

- Use the **default settings** for **Bend Allowance** and **Auto-Relief**.

- Click **OK**.

- Click OK when a message appears indicating the Auto-Relief-Cuts were added to some of the corners of the part.

Select the
Fixed face

- Zoom in to see the relief corners.

Relief Corner(s)

7. Viewing the Flat Pattern:

- From the Sheet Metal tool tab, click

the **Flatten** command .

- The Housing is
flattened. The
orientation of the
flattened view is based
on the Fixed face that was
specified in step number 6.

- Click **Flatten** again to fold it back.

- Click-off the **Edit Component** command.

8. Converting the 2nd component:

- Select the **Card Guide Left** as shown and click the

Edit Component command.

- From the Sheet Metal tool tab, click **Insert Bends** .

- Select the **Fixed face** as noted.

- Enter **.010"** for Bend Radius. Keep all other default parameters.

- Click **OK**.

- Click off the
Edit Component
command.

Select Fixed face

9. Converting the 3rd component:

- Select the **Card Guide Middle** as shown and click the **Edit-Component** command.

- From the Sheet Metal tool tab click **Insert Bends**.

- Select the **Fixed face** as noted.

- Enter **.010"** for Bend Radius and use the **default settings** for the Bend Allowance and K-Factor.

- Click **OK**.

- Click-off the **Edit Component** command.

Select Fixed face

10. Converting the 4th component:

- Select the **Card Guide Right** as shown and click the **Edit Component** command.

- From the Sheet Metal tool tab, click:

Insert Bends .

- Select the **Fixed face** as noted.

- Enter **.010"** for Bend Radius. Keep all other default parameters the same.

- Click **OK**.

Select Fixed face

- Click-off the **Edit Component** command.

11. Using the Hole-Series:

- The Hole Series is an Assembly-Feature; it is used to create a series of holes through the individual parts of an assembly.

- Unlike other assembly features, the holes are contained in the individual parts as externally referenced features. If you edit a hole series within an assembly, the individual parts are modified.

- From the Assembly tool tab, select **Assembly Features / Hole Series** .

Add 3 points

- From the FeatureManager, click: **Create New Hole** (arrow).

- The mouse cursor changes to the **Sketch Point** command.

- Create **3 points** approximately as shown.

- Each point is the center of a hole.

Along Z relation

4.000

4.000

1.500

.325

- Add an **ALONG Z** relation (Vertical) between the 3 points.

- Add the dimensions as indicated to fully define the positions of the points.

Continue...

- Click the **First Part** tab.

- Click the **Countersink** option.

- Select the following:

Standard: **Ansi Inch**

Type: **Flat Head Screw**

Size: **#4**

Fit: **Normal**

- Use the default settings for Custom Sizing.

- Click the **Middle Part** tab.

- Select the **Hole** button.

- Enable the check box:
  Auto Size based on Start Hole.

- Click the **Last Part** tab.

- Select the **Straight Tap** button.

- Enable **Auto Size Based On Start Hole**.

- Set Type to **Tapped Hole**.

- Set Size to **#4-40**.

- Set both End Conditions to **Through All**.

- Enable the checkbox: **With Thread Callout**.

- Click **OK**.

12. Using the Hole Wizard:

- Hole wizard is an Assembly Feature, which creates these types of holes:

  - Counterbore
  - Countersink
  - Hole
  - Straight Tap
  - Tapered Tap
  - Legacy

- From the Assembly tool tab, click **Assembly Features / Hole Wizard**.

- Select the **Countersink** option under the **Type** tab.

- Set the options **to match** the last 3 holes in step number 11.

- Switch to the **Positions** tab (arrow).

- Click approximately as shown to create **3 points**.

- Each point represents the center for that hole.

- Add the relation and dimensions as indicated.

- Click **OK**.

13. Verifying the two hole types:

- Although the 6 holes were created with 2 different hole options, they are exactly identical.

- Open the Card Guide Middle to verify that the holes are actually there on the part.

- The new feature in step 12 (propagate feature to parts) allows these Assembly Features to appear in the part mode as well.

14. Adding holes on the Card Guide Right:

- Repeat either step number 11 (Hole Series) or step number 12 (Hole Wizard) and create 3 more holes for the last Card Guide.

Add 3 more C'Sinks

- Use the same dimensions from the previous step to position the holes.

15. Adding the Smart Fasteners:

- Click **Smart Fasteners** 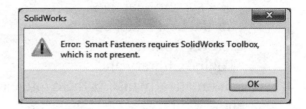 from the Assembly tool tab.

- An error message appears inquiring for SolidWorks Toolbox to be activated** (Requires SolidWorks Professional or SolidWorks Premium).

** To activate Toolbox, select the following:

- **Options / Add Ins**.

Enable Toolbox Library & Toolbox Utilities

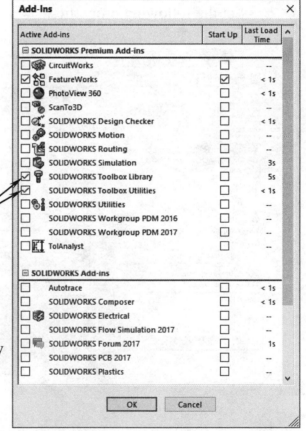

- Enable **SolidWorks Toolbox Library** and **SolidWorks Toolbox Utilities**.

- Click **OK** OK .

- Click the **Smart Fasteners** button once again.

- Another message pops up indicating that the Smart Fasteners calculation may take extra time; click **OK** OK .

- Select one of the Countersink holes from the graphics area.

- Click **Populate All** (arrow).

- The system searches for the best matched screws from its Toolbox library and automatically inserts them into each hole.

- Set the following properties:

    - Size: **#4-40**   Length: **.125in***   Drive Type: **Cross**

    - Thread Display: **Schematic**

    * *Change the screw length to .250in if adding a washer and a nut.*

- Click **OK**.

16. Creating an Exploded View:

Option 1

- Create an exploded view with all 4 parts
  shown in folded stage as shown.

- When an exploded view is created, SOLIDWORKS
  also creates an animated configuration, which
  can be played back and saved as an AVI
  file format.

Option 2

- Create a 2nd exploded view with all
  4 parts as shown in the Flatten view below.

NOTES:

- _Edit each component in order to switch from the
  Folded to flatten stage._

- _Configurations can also be used to capture
  the flat pattern of each component._

17. Saving your work:

- Click **File / Save As**.

- For the name of the file enter:
  SM_Assembly_Smart Fasteners.

- Click **Save**.

- Close all documents.

Adding Parts to the Toolbox Library

Customized parts or fasteners can be added to existing Toolbox folders. For parts that are stored in the shared library, administrators can control who can add or change parts by creating a Toolbox password and setting permissions for Toolbox functions for each user (see Toolbox Permission in the SOLIDWORKS Help section).

1. Starting the Toolbox Settings Utility:

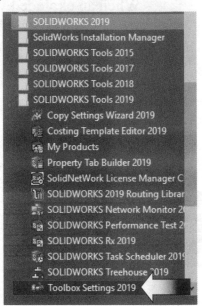

- From the Windows desktop select the following: **Start / All Programs / SOLIDWORKS 2019 / SOLIDWORKS Tools 2019/ Toolbox Settings** (arrow).

- Select the option number 2 (arrow): **Customize your hardware**.

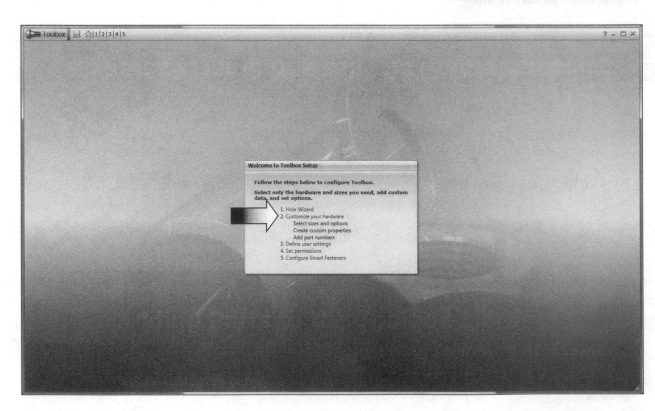

2. Adding a part to a folder:

- From the upper left side of the tree, select the following:

   ANSI Inch
   Bolts and Screws
   Countersunk Head

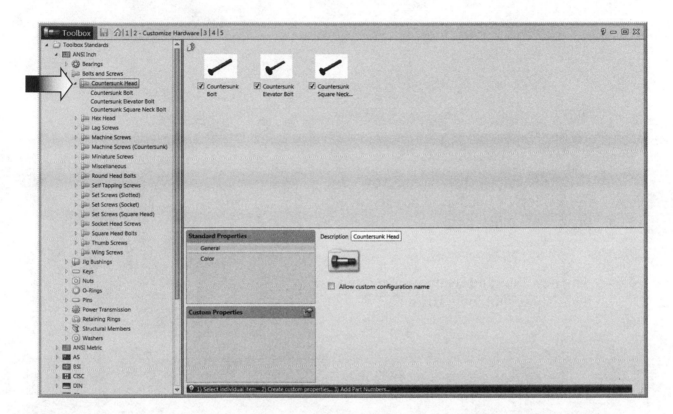

- In this folder there are 3 existing screw / bolt types: **Countersunk Bolt**,
   Counter-sunk Elevator Bolt, and **Countersunk Square Neck Bolt**.

- Right click the **Countersunk Head** folder and select **Add File** (arrow).

- Browse to the Training Files folder and select the document named:
C-Sink .250-20x1x0.75-Bolt.sldprt and open it.

- Click the floppy disk icon (arrow) on the top left of the dialog box to **save** the newly added part.

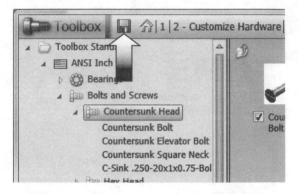

- The new part and its name appear in the display window (arrow).

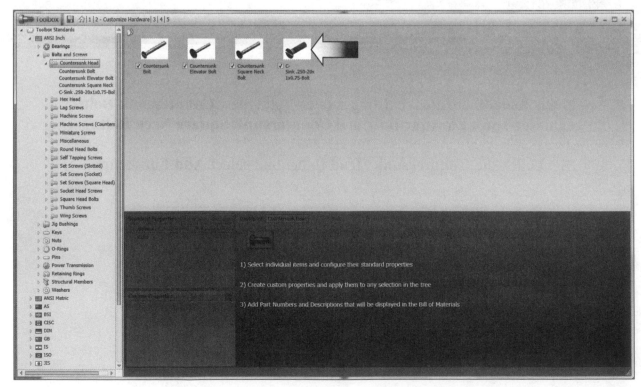

3. Activating Toolbox:

- Launch the SOLIDWORKS application.

- Select: **Options / Add-Ins** (arrow).

- Enable the 2 options:

    * **SOLIDWORKS Toolbox Library**
    * **SOLIDWORKS Toolbox Utilities**

- Click **OK**.

- From the **Task Pane** (on top right hand corner) click the **Design Library** folder and then expand the **Toolbox** folder (arrow).

NOTE: Toolbox is only available in SOLIDWORKS Professional and SOLIDWORKS Premium.

4. Locating the new part:

- Select the following under the Toolbox folder:

    * **ANSI Inch**
    * **Bolts and Screws**
    * **Countersunk Head**

- Locate the new part **C-Sink .250-.20x1x0.75-Bolt** on the lower right side of the task pane (arrow).

5. Viewing the new part:

- Right click on the new part and select:
  Configure Part.

NOTE:

- There is no configuration created for this custom screw; its length will have to be modified manually.

- Existing Toolbox parts will have several configurations such as Size, Length, Thread Display, etc., to choose from.

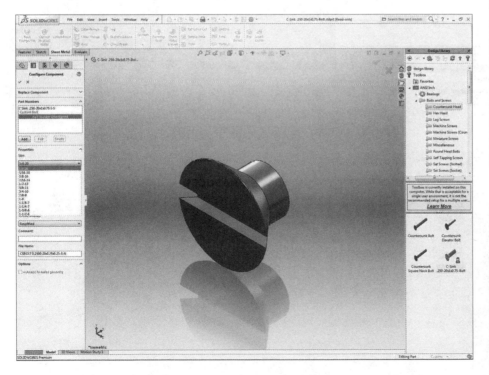

- The new Toolbox part appears in its own window.

6. Adding a Part Number and Description:

- Click the **Edit** button (arrow).

- Enter a new **Part Number** (if applicable).

- Enter **Custom Bolt** as Description.

- **Save** and close your documents.

CHAPTER 17

Advanced Weldments

Advanced Weldments
Weldments Platform

Weldments can be used to create a structure as a single multibody part. Either a 2D or 3D sketch can be used to define the framework, then create structural members containing groups of sketch segments.

Other Weldments tools can be used to add objects such as gussets and end caps. Weld gaps and beads can also be added to the model. Drawings can be made to document the design, including tables of cut materials, cut length, and weld bead totals.

Weldments are normally created in groups. A group is a collection of related segments in a structural member. A group is configured to affect all its segments without affecting other segments or groups in the structural member.

Types of groups are:
Contiguous: A continuous contour of segments joined end-to-end. You can control how the segments join to each other. The end point of the group can optionally connect to its beginning point.

Parallel: A discontinuous collection of parallel segments. Segments in the group cannot touch each other.

* You can define a group in a single plane or in multiple planes.
* A group can contain one or more segments.
* A structural member can contain one or more groups.
* After you define a group, you can operate on it as a single unit. Use the Structural Member PropertyManager to specify the corner treatment for the segments in the group.

Advanced Weldments
Weldments Platform

Dimensioning Standards: **ANSI**
Units: **INCHES** – 3 Decimals

Tools Needed:

3D	3D Sketch		Extruded Boss-Base
	Linear Pattern		Structural Member
	Trim/Extend		End Trim Type

1. Opening a part document:

- Click **File / Open**.

- Browse to the Training
  Folder and open a part
  document named:
  Advanced Weldments

2. Adding the Weldments toolbar:

- Right click one of the tool tabs and enable the **Weldments** toolbar.

- Select the **Structural Member** command 🔳.

- Set the following:

    * Standard:
      ANSI Inch

    * Type:
      Square Tube

    * Size:
      2 x 2 x 0.25

- You can define a group in a single plane or in multiple
  planes. After you define a group, you can operate on it
  as a single unit.

3. Adding the Structural Members:

- Start by selecting the **3 horizontal lines** on top as indicated.

- Group 1 is created automatically. This is a Parallel group.

- Click the **New Group** button (arrow) to start a new group.

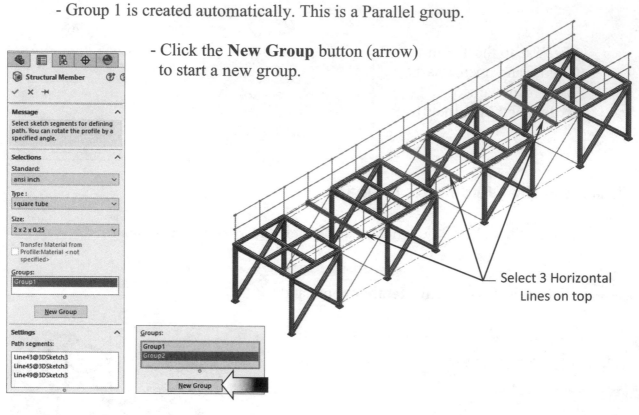

Select 3 Horizontal Lines on top

- For Group 2, select the **3 vertical lines** also from the top.

- Group 2 is created. This is also a Parallel group.

- Typically, we can continue to select other groups, but we will stop here and take a look at how the tubes are trimmed.

Select 3 Vertical Lines on top

- Click **OK**.

- Zoom in on the first two sections of the platform frame.

- Notice the new tubes were trimmed automatically.

Trimmed automatically

4. Adding more Structural Members:

- Click the **Structural Member** command ⬛ again.

- Keep all parameters the same as the last group and select the **6 parallel lines** on both sides as noted.

- The new Group 1 is created. This is also a Parallel group.

Select 6 parallel lines on both sides

- Click the **New Group** button to start a new group.

- For Group 2, leave all settings the same as the last group and select the **6 parallel lines** on both sides as indicated in the image.

- Group 2 is created. Similar to the other groups, we were selecting only the lines that are parallel to each other so that later on we can modify or work with each group separately.

Select 6 parallel lines on both sides

- Click the **New Group** button.

- For the new Group 2, select the **10 horizontal lines** for the rails.

- Group 2 is created but the weldment platform is becoming very busy, making it hard to see the other lines.

- For clarity, we will hide the Sketch2 to make the other lines easier to see.

- Click **OK** to exit the command.

- Click the **Sketch2** on the Feature Tree and select **Hide**.

Select 4 LONG horizontal lines and 6 SHORT horizontal lines

5. Adding the vertical Structural Members to the rails:

- Click the **Structural Member** command again.

- For the new Group 1, select the **30 vertical lines** as indicated.

- The new Group 1 is created with all parallel lines in it. Again, by grouping the lines, they become one unit. Editing will be much easier in the future.

Select 30 vertical lines
on both sides

- Click **OK** to exit the command.

- Click on the **Sketch3** and select **Hide** (arrow).

- Compare your model with the image shown here.

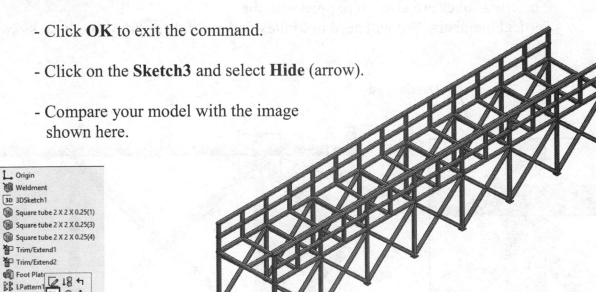

6. Viewing the overlapped areas:

- Zoom in on the area shown in the enlarged view to see the overlapped issues.

Overlapped

- The route lines of these tubes were drawn to the center of the horizontal tubes. We will need to trim both tubes on the left and the right sides.

- The cross-tubes are also overlapped with the vertical members. We will need to trim/extend all of them, front and back sides.

Overlapped

7. Trimming the overlapped areas:

- Click the **Trim/Extent** command 🗗.

- Use the default **End Trim** option 📦.

Trimming Boundary:
select 4 horizontal tubes

Bodies to be trimmed:
select 30 vertical tubes

- For Bodies to be Trimmed, select the **30 vertical members** as indicated.

- For Trimming Boundary, click the **Bodies** button and select the **4 horizontal tubes** as noted.

- The preview graphics shows the vertical members are trimmed to match the faces of the horizontal bodies.

- Click **OK** to exit the Trim.

8. Trimming the cross-members:

- Click the **Trim/Extent** command .

- Use the default **End Trim** option.

- Change to the **Right** orientation (Control + 4).

- For Bodies to be Trimmed, select the **Cross-member** indicated.

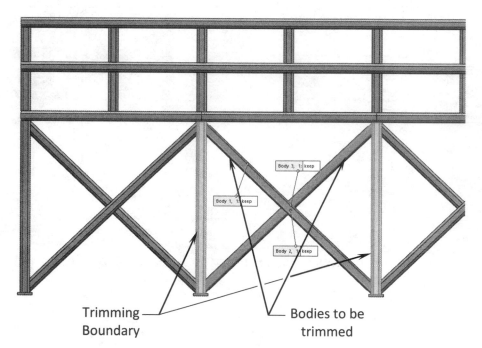

Trimming Boundary

Bodies to be trimmed

- For Trimming Boundary, select the **vertical member** as noted.

- The preview graphics shows the cross-member is trimmed to the right face of the vertical tube (circled).

- Click **OK** to exit the trim command.

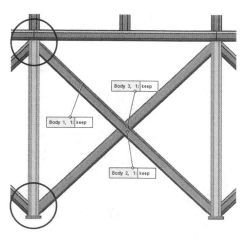

9. Trimming the other structural members:

- There are several other structural members throughout the platform that still need to be trimmed (circles).

- Change to the **Right** orientation (Control + 4).

- Click the **Trim/Extend** command and trim the tubes shown in the circles.

10. Updating the Cut list:

- SOLIDWORKS updates the cut list automatically when new features are added, such as extruded foot plates or 3D bounding boxes, existing features are edited, or the model is rebuilt. The model's custom properties and internal supporting data are also updated, preventing custom property errors.

- This Icon in front of the Cut List [icon] indicates the Cut List needs updating.

- This Icon in front of the Cut List [icon] indicates the Cut List is up to date.

- Right click the **Cut List** icon and select **Update** (arrow). The Cut List items that were created as groups are automatically sorted and placed in separate folders.

- To view the Cut List details, right click on one of the Cut List-Items, and select: **Properties** (arrow).

- Select one of the items such as Angle, Length, Description, etc. to see the details of the item.

- Click **OK** to exit the Cut List Properties.

11. Saving your work:

- Select File / **Save As**.

- Enter **Weldments Platform.sldprt** for the name of the file.

- Click **Save**.

Using Weldments – Structural Members

The options in Weldments allow you to develop a weldment structure as a single multibody part. The basic framework is defined using 2D or a 3D sketch, and then structural members like square or round Tubes are added by sweeping the tube profile along the framework. Gussets, end caps, and weld-beads can also be added using the tools on the **Weldments** toolbar.

1. Opening an existing document:

- Open a file named **Weldment Frame** from the Training Files folder.

2. Enabling the Weldments toolbar:

- Right click the **Evaluate** tab and select the **Weldments** toolbar from the list (arrow).

- Click the **Weldments button** from the weldment toolbar.

- A Weldment Feature appears on the feature tree with a Weldment Cut List (arrows), which indicates the items from the model to include in this cut list.

- A single 3D Sketch is created for the purpose of this exercise. Multiple sketches (2D & 3D) can be used to work with weldments.

3. Adding Structural Members:

- Click the **Structural Member** button from the Weldments toolbar.

- Select the following:

   * **Ansi Inch.**
   * **Square Tube**
   * **4 X 4 X 0.25**

- Click the **4 lines** on the top of the frame.

- Select the **MITER** under Apply Corner-Treatment.

- Click **OK**.

- By default, the profile of the tube is automatically centered on the end of each line.

- Try out all 3 options for corner treatments: End Miter, End Butt1, and End Butt2.

- Switch back to the End Miter option when finished.

Select the 4 upper lines

1st Group

End Miter	**End Butt1**	**End Butt2**
	(Overlapped)	(Under-lapped)

4. Adding Structural Members to Contiguous Groups*:

- Repeat the previous step and add another 4 square tubes to the 2nd group as shown.

- Use these same settings:

    * **Ansi Inch.**
    * **Square Tube**
    * **4 X 4 X 0.25**

2nd Group

Contiguous Group

* A group is a collection of related segments in a structural member. There are 2 types of groups. One is called Contiguous, where a continuous contour of segments is joined end-to-end. The other is called Parallel, which includes a discontinuous collection of parallel segments. Segments in the group cannot touch each other.

3rd Group

Contiguous Group

- Repeat the same step for the 3rd group.

- Follow the same procedure and add the same size tubing to the vertical members as noted for the 4th group.

Note: *Select the exact same Vertical tubes on both sides (total of 12).*

4th Group
(both sides)

5. Adding Structural Members to the Parallel Groups:

- Repeat the previous step and add the same structural members to the 5th group as indicated.

5th Group
(both sides)

Parallel Group

 Groups

You can define a group in a single plane or in multiple planes. A 3D sketch is best suited for weldment designs since all entities can be drawn and controlled in the same sketch.

- Create the same type of structural members for the 6th and 7th group, which has only 2 lines in each group...

6th Group
rotate 30 deg.

Parallel Group

- Rotate the profile to **30 deg.** for the 6thth group and **60 deg.** for the 7th group.

- There are several over-lapped areas that need trimming; we will look into that in the next steps.

Parallel Group

7th Group
rotate 60 deg.

6. Hiding the 3D sketch:

- Right click on one of the lines
  in the 3D sketch and select **Hide** .

- Notice the overlapping areas in the enlarged view below?
  For practice purposes, we will learn to use different trim
  options to cut the tubes to their exact lengths and angles.

Overlapped

Overlapped

7. Trimming the Structural Members:

- Click **Trim/Extend** on the Weldments toolbar.

Select 4 Bodies
To Be Trimmed

- For **Corner Type**, use the default **End Trim** (arrow).

- For **Bodies To Be Trimmed**, select the 4 vertical tubes as shown.

- For **Trimming Boundary**, select the **Body** option (arrow).

- For **Trimming Bodies**, select the **6 horizontal tubes** as indicated.

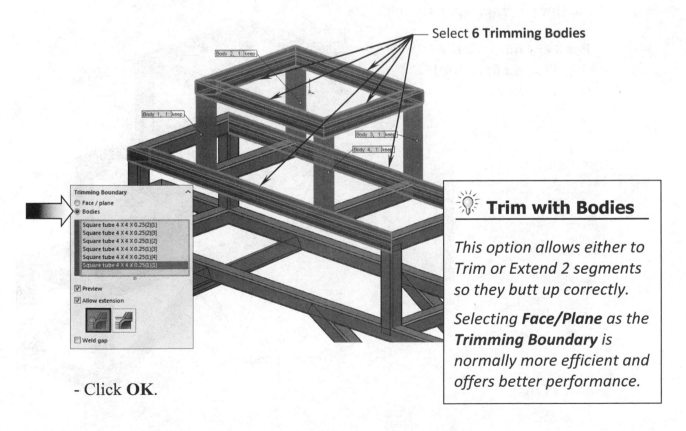

Select **6 Trimming Bodies**

💡 Trim with Bodies

This option allows either to Trim or Extend 2 segments so they butt up correctly.

*Selecting **Face/Plane** as the **Trimming Boundary** is normally more efficient and offers better performance.*

- Click **OK**.

8. Trimming the Parallel Groups:

- Select the **Trim/Extend** command once again from the Weldments toolbar.

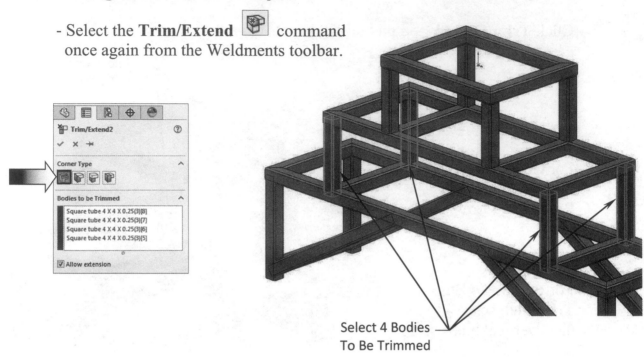

Select 4 Bodies
To Be Trimmed

- For **Bodies To Be Trimmed**, select the next 4 vertical tubes as noted.

- For **Trimming Boundary**, select the **Body** option again (arrow).

- For **Trimming Bodies**, select the **6 horizontal tubes** as shown.

Select **6
Trimming Bodies**

- Click **OK**.

9. Trimming the next Parallel Groups:

- Select the **Trim/Extend** command from the Weldments toolbar.

- For **Bodies To Be Trimmed**, select the **4 vertical tubes** on the bottom as noted.

- For **Trimming Boundary**, click the **Body** option (arrow).

Select 4 Bodies To Be Trimmed

- For **Trimming Bodies**, select the **4 horizontal tubes** as shown.

Select 4 Trimming Bodies

- Click **OK**.

10. More Trimming:

- Select the **Trim/Extend** command from the Weldments toolbar.

Select 6 Bodies
To Be Trimmed

- For **Bodies To Be Trimmed** select the **6 short horizontal tubes** as indicated.

- For **Trimming Boundary**, click the **Body** option (Arrow).

- For **Trimming Bodies**, select the **8 structural members** as indicated.

Select 8
Trimming Bodies

- Click **OK**.

11. Trimming with Face/Plane:

- Select the **Trim/Extend** 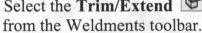 command from the Weldments toolbar.

Select 2
Bodies To
Be Trimmed

- For **Bodies To Be Trimmed** select the 2 structural members as noted.

- For **Trimming Boundary**, click the **Face / Plane** option (arrow).

- Select the planar surface as noted, for **Trimming Bodies**.

Select Face
for Trimming
Boundary

💡 Trim with Face/Plane

This option allows a planar face(s) as a trimming boundary to trim one or more solid bodies.

*Selecting **Face/Plane** as the **Trimming Boundary** is normally more efficient and offers better performance.*

- Click **OK**.

12. More Trimming with Face/Plane:

- Select the **Trim/Extend** command from the Weldments toolbar.

Select 2 Bodies
To Be Trimmed

- For **Bodies To Be Trimmed** select the **2 structural members** as indicated.

- For **Trimming Boundary**, click the **Face / Plane** option (Arrow).

- For **Trimming Bodies**, select the planar surface on the back of the vertical tube as noted.

Body 2, 2: keep

Body 1, 1: keep

Select Face
(rotate around)
for Trimming
Boundary

Body 2, 1: discard

Body 1, 2: discard

- Click **OK**.

13. Trimming the last 4 structural members:

- Select the **Trim/Extend** command
  from the weldment toolbar.

Select 4 Bodies
To Be Trimmed

- For **Bodies To Be Trimmed** select the **4 structural members** as shown.

- For **Trimming Boundary**,
  click the **Body** option
  (arrow).

- For **Trimming Bodies**,
  select the **2 horizontal
  tubes** as noted.

Body 2, 1: keep

Body 1, 1: keep

Body 3, 1: keep

Body 4, 1: keep

Select **2**
Trimming Bodies

- Click **OK**.

(Front View)

14. Adding the foot pads:

- Insert a <u>new sketch</u> on the
  bottom surface of one of
  the 4 legs.

- Sketch the profile as shown.

- The 4 circles are concentric
  with the corner radiuses.

- Add the dimensions and relations needed to fully define the sketch.

- Mirror the sketch to make a total of 4 foot pads.

<u>*Note:*</u>
*Add a couple of centerlines
as shown prior to making
the mirror.*

Add centerlines
to mirror

- Since the 4 foot pads have the same thickness,
  we can extrude them at the same time.

- Click **Extruded Boss/ Base**.

- Enter the following:

    * Type: **Blind**

    * Depth: **1.000**

- Click **OK**.

15. Adding the Gussets:

- Rotate and zoom to an orientation that looks similar to the view below.

- Click the **Gusset** command.

- Enter the following:

- For **Supporting Faces**, select the **2 faces** as noted.

    * Distance1: **4.00 in.**

    * Distance2: **4.00 in.**

    * Distance3: **.500 in.**

    * Thickness: **.500 in.** (Both Sides)

    * Location: **Midpoint**

- Click **OK**.

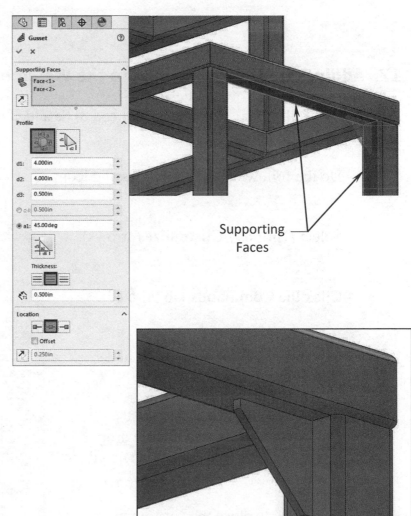

Supporting Faces

16. Adding more Gussets:

- Repeat the step 15 and add a gusset to each corner of the frame.

- Next, we are going to add the weld beads around the gussets. Weld beads can be added as full length, intermittent, or staggered fillet weld beads between any intersecting weldment entities such as structural members, plate weldments, or gussets.

Add more Gussets

17. Adding the Fillet-Beat icon to the Weldments toolbar:

- The **Fillet Beat** icon may need to be added to the Weldments toolbar.

- Do the following to add the missing icon:

- Select **Options / Customize** (arrows).

- Click the **Commands** tab (circled).

- Select the **Weldments** option under **Categories** (arrow).

- Drag the **Fillet Bead** icon and drop it onto the Weldments toolbar as noted.

- Click **OK** to close the Customize dialog.

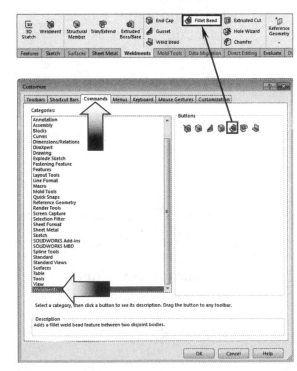

18. Adding the Fillet Beads:

- Click **Fillet Bead** on the Weldments toolbar.

- From the Weld Bead properties tree, enter the following:

* Bead Type: **Full Length**

* Fillet Size: **.250 in.**

* Tangent Prop: **Enabled**

* Face Set1: **Select the 2 faces** as noted.

Select **2 Faces** (front & back) for Face Set1

- For **Face Set2**, select the next **2 faces** as indicated.

Intersecting Edges:
Highlights edges where
Face Set1 *and* ***Face-Set2*** *intersects.*

*(You can also right-click an edge and select **Delete** to remove from the weld bead.)*

Select **2 Faces**

- Enable the **Other Side** checkbox and apply the <u>same settings</u> to the back end of the gusset.

- Different bead type or fillet size can be added to the other side, but we are going to use the same settings as the first side.

- Click **OK**.

- A bead call out is added automatically (see example below).

    Example: *0.25   = Length of the leg of the fillet bead.*
                         0.375 = Length of each bead segment.
                         0.7   = Distance between the start of each bead.

Intermittent or Staggered

19. Adding more Fillet Beads:

- Repeat step 17 and add a set
  of fillet beads to each gusset
  that was created earlier.

- When adding the fillet beads,
  try using the different types
  of beads: Full Length,
  Intermittent, and Staggered
  to see the different results
  and callouts.

Add more
Fillet Beads

20. Viewing the Weldment Cut List:

- Locate the **Cut List** on the FeatureManager
  tree and click the Plus (+) sign to expand.

- The Cut List needs to be updated
  every time something is added
  to the model.

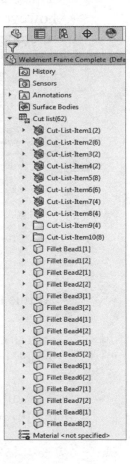

- The icon in front of the cut list
  indicates that it needs updating and

  the icon indicates the list is up to date.

- The current list displays all items in the order that they
  were created. We will update the list in the next step.

21. Updating the Cut List:

- Right click on the cut list and select **Update** (arrow).

- Notice the option **Automatic** is on by default? This option organizes all of the weldment entities in the cut list for the new weldment parts.

- Although the cut list is generated automatically, you can manually specify when to update the cut list in a weldment part document.

- This enables you to make many changes, and then update the cut list once; however, the cut list updates automatically when you open a drawing that references the list.

- At this time the cut list icon is changed to which indicates the cut list is now up to date.

- The cut list <u>Table</u> can now be viewed by accessing the Properties of one of the item folders.

22. Creating a drawing (OPTIONAL):

- A drawing that includes the cut list can be generated. (Refer to the Part-1 Basic Tools textbook for more information on how to create a detail drawing in SOLIDWORKS 2019.)

23. Saving your work:

- Click **File / Save As**.

- Enter **Weldments Frame** for the name of the file.

- Click **Save**.

- Replace the existing file when prompted.

Optional:

To create the exploded view similar to the one shown here:
Either use the Move/Copy command (Insert / Features/Move-Copy)
or use the Exploded View command (Insert /Exploded View).

CHAPTER 18

Creating a Core and Cavity

Creating a Core and Cavity
Linear Parting Lines

A mold is normally designed in SOLIDWORKS using a sequence of intergraded tools that control the mold creation process. Using the finished model, these mold tools can be used to analyze and correct deficiencies in the part.

The process usually follows these steps: Draft analysis, Undercut Detection, Parting Lines, Shut-Off Surfaces, Parting Surfaces, Interlock Surfaces (Ruled Surfaces), and Tooling Split.

The Parting Lines lie along the edge of the molded part, between the core and the cavity surfaces. They are used to create the Parting Surfaces and to separate the surfaces.

The Shut-Off Surfaces are created after the Parting Lines. A shut-off surface closes up a through hole by creating a surface patch along the Edges that form a continuous loop, or a parting line you previously created, to define a loop.

After the Parting Lines and the Shut-Off Surfaces are determined, the Parting Surfaces are created. The Parting Surfaces extrude from the parting lines and are used to separate the mold cavity from the core.

After a parting surface is defined, the Tooling Split tool is used to create the core and cavity blocks for the model. To create a tooling split, at least three surface bodies are needed in the Surface Bodies folder: a Core, and a Cavity and a Parting surface.

With most mold parts, the interlock surfaces need to be created. The interlock surfaces help prevent the core and cavity blocks from shifting, and are located along the perimeter of the parting surfaces. Usually they have a 5-degree taper.

Creating a Core and Cavity
Linear Parting Lines

Dimensioning Standards: **ANSI**

Units: **INCHES** – 3 Decimals

Tools Needed:

 Parting Lines

 Parting Surfaces

 Shut-Off Surfaces

 Tooling Split

 Planes

 2D Sketch

1. Opening an existing Parasolid document:

- Select **File / Open**.

- Browse to the Training Files folder, change the Files of Type to **Parasolid**.

- Select **Remote Control.x_b** and click **Open**.

Scale

The Parasolid part has already been scaled to 1.05% about its Centroid.

- Click **NO** on the Import Diagnostic dialog and close it.

2. Creating the Parting Lines:

- Right click the Evaluate tab and enable the **Mold Tools** option.

- Click **Parting Line** on the Mold Tools toolbar.

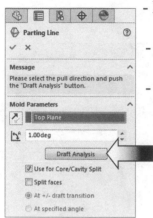

- From the Feature tree, select the **Top** plane for Direction of Pull.

- Enter **1deg** for Draft Angle.

- Click **Draft Analysis**.

Straddle Faces

Negative Drafts

No Drafts

Positive Drafts

- SOLIDWORKS automatically selects the lower edges of the part where the red and the green faces meet up, and places them in the Parting Lines section.

Straddle Faces

* Displays any faces that contain both positive and negative types of draft.
* Typically, these are the faces that require creating a split line.

- Click **OK**.

3. Creating the Shut-Off Surfaces:

- Click 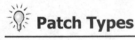 or select **Insert / Molds / Shut-Off Surfaces**.

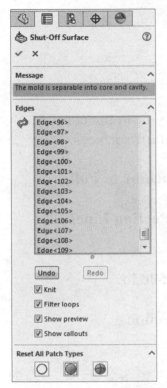

- The system automatically selects the edges of all through openings and labels them as Loop/Contacts.

- Keep the Patch Type at **Contact**.

Resets all through hole surface patches to one of the 3 settings.

— All No-Fill

— All Contact

— All Tangent

💡 **Patch Types**

Only one Shut-Off Surface feature is allowed in a model. Therefore, within the one feature, you must assign a fill type of **Contact**, **Tangent**, or **No Fill** to every hole.

- Click **OK**.

4. Creating the Parting Surfaces:

- Click 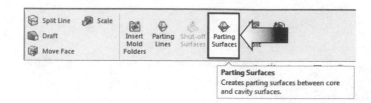 or select **Insert / Molds / Parting Surfaces**.

Parting Surfaces
Creates parting surfaces between core and cavity surfaces.

- The Parting surfaces split the mold cavity from the core. It gets created after parting lines and shut off surfaces.

- In the Mold Parameters, select **Perpendicular to Pull**.

- In the Parting Line selection, select the **Parting Line1** from the FeatureManager Tree.

- In the Parting Surface selection, enter **1.500 in**.

- Select **Sharp Edges** under Smoothing section.

— Smooth Edges

— Sharp Edges
(A higher value creates a smoother transition between adjacent edges)

- Click **OK**.

5. Sketching the profile of the mold-blocks:

- Select the <u>Top</u> plane and open a new sketch .

- Sketch a Center Rectangle approximately as shown.

- Add the width and the height dimensions .

- The sketch should be fully defined at this point.

- **<u>Exit</u>** the Sketch .

(The next step is to create the upper and lower blocks using the Tooling Split command. This is only available when the Sketch is off.)

6. Creating the Tooling Split:

- Click or select:
  Insert / Molds / Tooling Split.

- In the Block Size selection, enter:

    1.500 in for upper block.
    1.50 in for lower block.

Upper block thickness

Lower block thickness

- The Cavity and the parting surfaces options should already be filled.

- Click **OK**.

* The Interlock Surface surrounds the perimeter of the parting lines in a slight tapered direction; it helps seal the mold to prevent resins from leaking, prevents shifts, and maintains alignment between the tooling entities.

7. Hiding the Solid Bodies:

- From the FeatureManager Tree, expand the Surface Bodies folder. There are 3 groups of surfaces in this folder.

- Click the **Surface Bodies** folder and select **Hide** .

- The 3 surfaces that were created in the previous steps are temporarily removed from the graphics display.

- Change to the Wireframe mode to see the inside details of the blocks.

- Right click on the blue parting lines and hide it (arrow).

8. Saving the bodies as part files:

- Expand the **Solid Bodies** folder, right click on **Tooling Split [1]** and select: **Insert into New Part**.

- Click the **Browse** button under the Transfer section.

- Enter: **Core_Remote_Control** for the name of the file.

- Click **Save**.

- Hold the **Control** key and push the **Tab** key to switch back to the main part.

- Repeat the last step to save the **Tooling Split[2]**, enter: **Cavity_Remote_Control** for the name of the 2nd block.

9. Separating the 2 blocks:

- Select **Insert / Feature / Move-Copy**.

- Select the upper block in the graphics.

- Click the vertical **Green** arrow to define the direction.

- Under the Translate section, enter **4.00in** and press **Enter**.

- Click **OK**.

- The upper block moves 4 inches upward from its original position.

Select the **Y** (vertical arrow) and enter the distance (arrow)

- Repeat the same step to move the lower block downward (use -4.00" for distance).

10. Saving your work:

- Save a copy of your work as **Remote Control Tooling**.

11. Optional:

- Start a new Assembly document and assemble the 3 components.

- Create an Assembly Exploded View as a separate configuration.

- Add Injector hole.

- Ejector holes.

- Alignment Pins.

- Make copies of the components and create an exploded view as shown.

Exercise: Linear Parting Lines

1. Opening a part document:

- Open the part document named: **Ice Cube Tray.sldprt** from the training files folder.

2. Enabling the Mold Tools:

- Right click one of the tool tabs and select the **Mold Tools** (arrow).

- A material has already been assigned to the part (PP Copolymer) but a new material will be assigned to the mold blocks after they are created.

3. Applying Scale:

- Switch to the **Mold Tools** tab.

- Click the **Scale** command and enter **1.02** (2% larger).

- Use the default **Centroid** option.

- Click **OK**.

4. Creating the parting line:

- Click the **Parting Lines** command.

- For Direction of Pull, select the **Top** plane.

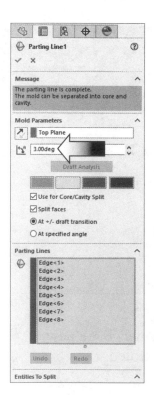

- For Draft Angle, enter **3.00deg**.

- Enable the **Use for Core/Cavity Split** checkbox.

- Enable the **Split-Faces** box.

- Zoom in on one of the corners, right click on an edge <u>between</u> the Green and the Red surfaces, pick **Select Tangency**.

- Click **OK**.

Right click an edge between the Green & Red surfaces

5. Creating the parting surface:

- Click the **Parting Surfaces** command.

Parting Surfaces
Creates parting surfaces between core and cavity surfaces.

- Under the Mold Parameters, select the **Perpendicular to Pull** option (arrow).

- For Parting Surface, enter **2.500in** for Distance (arrow).

- Enable the **Knit All Surfaces** checkbox.

- Click **OK**.

6. Sketching the mold block:

- Open a new sketch on the Parting Surface.

- Sketch a **Center-Rectangle** that is centered on the Origin.

- Add the dimensions shown to fully define the sketch.

- **Exit** the sketch.

7. Creating the tooling split:

- Switch back to the **Mold Tools** tab.

- Select the new sketch from the FeatureManager tree and click **Tooling Split**.

- For Upper Block Size, enter **1.000in**.

- For Lower Block Size, enter **2.000in**.

- The reference surfaces are automatically placed in their correct locations.

- Click **OK**.

8. Hiding the reference surfaces:

- Locate the **Surface Bodies** folder near the top of the
  FeatureManager tree.

- Click the Surface Bodies folder and select **Hide** (arrow).

9. Creating an exploded view:

- Select **Insert / Exploded View** (arrow).

- Select the **upper mold block** and drag the **Y** arrowhead upwards
  approximately about 6 inches.

Drag the Y
arrowhead
upward

- The Explode Step1 is created and stored under the Explode Step section.

- Select the **lower mold block** and drag the **Y** arrowhead downwards, approximately 6.5 inches.

- The Explode Step2 is created.

- Click **OK**.

10. Assigning materials:

- Expand the **Solid Bodies** folder.

Drag the Y arrowhead downward

- Right click the Body-Move/Copy1 (the upper mold block) and select: **Change Transparency**.

- Right click the same body and change the material to **Plain Carbon Steel**.

- Assign the same material to the lower mold block.

- Assign **PP Copolymer** to the plastic part.

11. Saving your work:

- Click File / Save As.

- Enter **Ice Cube Tray (Completed).sldprt** for the file name and click **Save**.

Questions for Review

Creating a Core and Cavity

1. Using the finished model, the mold tools can be used to analyze and correct the deficiencies such as undercuts, draft angles, shut-off surfaces, etc.
   a. True
   b. False

2. The Parting Lines are used to create the Parting Surfaces and to separate the surfaces.
   a. True
   b. False

3. A shut-off surface closes up a through hole by creating a surface patch along the edges that form a continuous loop.
   a. True
   b. False

4. The Parting Surfaces extrude from the parting lines and are used to separate the mold cavity from the core.
   a. True
   b. False

5. To create a tooling split, what surface bodies are needed for this operation?
   a. The Core
   b. The Cavity
   c. The Parting Surface
   d. All of the above

6. The Interlock surfaces help prevent the core and cavity blocks from shifting and are located along the perimeter of the parting surfaces.
   a. True
   b. False

7. The solid bodies can be hidden or shown just like any other features in SOLIDWORKS.
   a. True
   b. False

7. TRUE

5. D   6. TRUE

3. TRUE   4. TRUE

1. TRUE   2. TRUE

CHAPTER 19

Non-Planar Parting Lines

Non-Planar Parting Lines
Mold-Tooling Design

Using SOLIDWORKS a mold is created by following a sequence of integrated tools that control the mold creation process.

The mold tools are used to analyze and correct deficiencies such as draft angles or undercuts with the plastic models to be molded.

The mold tools span from initial analysis to creating the tooling split. The result of the tooling split is a multibody part containing separate bodies for the molded part, the core, and the cavity, plus other optional bodies such as side cores.

The multibody part file maintains your design intent in one convenient location. Changes to the molded part are automatically reflected in the tooling bodies.

The mold design process is as follows:

* **Draft Analysis:** Examines the faces of the model for sufficient draft, to ensure that the part ejects properly from the tooling.
* **Undercut Analysis:** Identifies trapped areas that prevent the part from ejecting.
* **Parting Line Analysis:** Analyzes transitions between positive and negative draft to visualize and optimize possible parting lines.
* **Parting Lines:** Creates a parting line from which you create a parting surface.
* **Shut-off Surfaces:** Creates surface patches to close up through holes in the molded part.
* **Parting Surfaces:** Extrude from the parting line to separate mold cavity from core. You can also use a parting surface to create an interlock surface.
* **Ruled Surface:** Adds draft to surfaces on imported models. You can also use the Ruled Surface tool to create an interlock surface.
* **Tooling Split:** Creates the core and cavity bodies, based on the steps followed earlier.

Non-Planar Parting Lines
Mold-Tooling Design

Dimensioning Standards: **ANSI**
Units: **INCHES** – 3 Decimals

Tools Needed:

Parting Lines	Tooling Split	Planar Surface
Shut-Off Surfaces	Ruled Surface	Knit Surface
Parting Surfaces	Filled Surface	Trim Surface

1. Opening an existing part document:

- Click **File / Open**.

- Browse to the Training Files folder, locate and open a part document named: **Receiver.sldprt**.

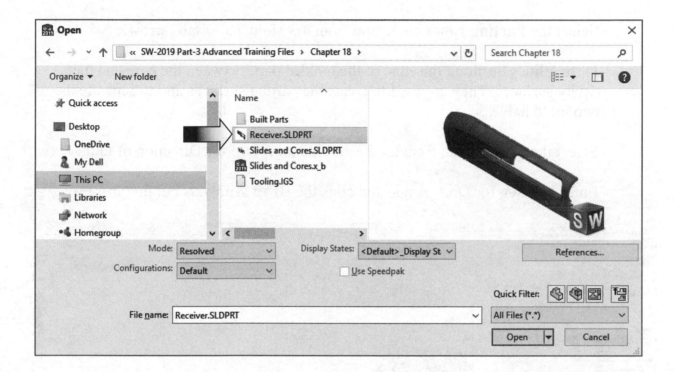

2. Enabling the Mold Tools toolbar:

- Right click on one of the existing tool tabs and enable the **Mold Tools** checkbox (arrows).

- For clarity, only keep the following toolbars enabled:

  * **Features**
  * **Sketch**
  * **Mold Tools**
  * **Evaluate**

- Disable the other toolbars.

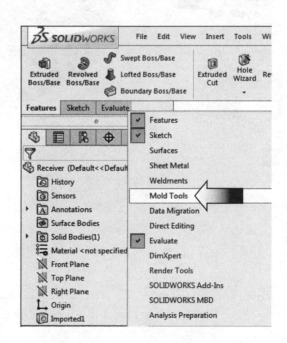

<u>**NOTE:**</u> *This model has already been scaled to accommodate mold shrinkage. The draft angles also have been added to all features. The next step is to add the parting lines.*

3. Creating the Parting Lines:

- Switch to the new **Mold Tools** tab (arrow).

- Select the **Parting Lines** command from the Mold Tools tab (arrow).

- Parting lines lie along the edge of the molded part, between the core and the cavity surfaces. They are used to create the parting surfaces and to separate the two mold halves.

- Select the **Front** plane from the FeatureManager tree for Direction of Pull.

- Enter **1.00deg** for Draft Angle and click the **Draft Analysis** button (arrow).

- SOLIDWORKS automatically selects the edges of the model that border the two halves of the mold. The parting lines will be used to separate the surfaces between the core and the cavity.

- The **Green** surfaces on the model represent the positive draft surfaces on the Cavity half, and the Red surfaces are negative draft surfaces on the Core half.

- Click **OK**.

4. Creating the Shut-Off Surfaces:

- A shut-off surface closes up a through hole by creating a surface patch along the edges that forms a continuous loop, or a parting line you previously created to define a loop.

- Click the **Shut-Off Surfaces** command on the Mold Tools tab (arrow).

Shut-Off Surface —

- SOLIDWORKS searches for any through holes and automatically creates a surface patch along the edges that form a continuous loop.

- A "Green Message" appears on the Feature tree
  indicating the mold is separable into core and cavity.

- Enable the **Knit** checkbox (arrow).

- Click **OK**.

5. Creating the Parting Surfaces:

- Parting surfaces are used
  to separate the mold cavity
  from the core. They must be created right after the Shut-Off Surfaces.
  (The Shut-Off Surface is hidden for clarity.)

Parting Surfaces ⎯

- Select the **Perpendicular to Pull** option (arrow).

- Enter **1.00in** for **Distance** (arrow).

- Use the default **Sharp*** option for smoothness (arrow).

Sharp*

Smooth

6. Creating a Ruled Surface:

- To help prevent the core and cavity blocks from shifting, you can add an interlock surface. It is created along the perimeter of parting surfaces prior to inserting a tooling split in a mold part. The interlock surface can be created manually or automatically.

- Because of the sudden changes in the parting surface geometry, the interlock surfaces will need to be created manually.

- The Ruled Surface command is used to create the tapered surfaces that form the interlocks.

- Select the **Rule Surface** from the Mold Tools tab (arrow).

Interlock Surfaces

- Click the **Taper to Vector** radio button (arrow).

- Enter **1.50in** for **Distance**.

- Select the **Front** plane for **Reference Vector**.

Skip these edges

- Select the edges along
  the Parting Surfaces
  and skip the ones
  as indicated.

- Double check your
  Ruled Surfaces
  against the
  ones shown
  here.

- Click **OK**.

Skip these edges

7. Creating the patches:

- We will use a combination of Lofted Surface and Filled Surface commands to create the patches and fill the openings in the ruled surfaces.

- Right click on one of the tool tabs and enable the **Surfaces** toolbar (arrow).

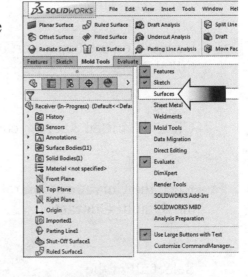

- Select the **Lofted Surface** from the Surfaces tool tab.

- Rotate the model to a position similar to the one shown here.

- Select the **2 edges** of the opening as noted.

Select 2 edges

- A preview of a lofted surface appears filling the lower portion of the opening.

- Click **OK**.

8. Patching with Filled Surface:

- The Filled Surface command creates a surface patch with any number of sides, within a boundary defined by existing model edges, sketches, or curves, including composite curves.

- Zoom in on the upper portion of the opening. We will patch it up with the Filled Surface command.

- Click the **Filled Surface** command from the Surfaces toolbar.

- Keep the Curvature Control at the default: **Contact**.

- Select all of the remaining edges on the upper portion of the same opening.

Select all edges —

- A preview of a filled surface appears filling the upper portion of the opening.

Filled surface —

- Click **OK**.

9. Patching all openings:

- Repeat the steps 7 and 8 and patch / fill the rest of the openings in the model.

Lofted surface
(left side)

Filled surface
(left side)

Lofted surface
(right side)

Lofted surface
(right side)

- The Filled Surface command requires the boundary to be closed in order to fill or patch that area. The Curve Through Reference Points command can be used to create the missing entities in those openings.

Opening

- From the **Surfaces** toolbar, select **Curves / Curve Through Reference Points**.

- Select the **2 vertices** as indicated and the preview of a curve appears.

- Click **OK**.

Curve Through 2 Reference Points

Filled surface

- Continue with creating the patches to fill the rest of the openings.

Lofted surface

Filled surface

- After all openings are patched, expand the Surface Bodies folder and <u>show all surfaces</u>.

10. Knitting the surfaces:

- To help select multiple surfaces more easily, we will need to knit all of the ruled surfaces and the patched surfaces together as one surface.

- Click the **Knit Surface** command (arrow).

- Expand the <u>Surface Bodies</u> folder and select all surfaces inside of the <u>Parting Surface</u> folder (arrow).

- Click **OK**.

11. Creating a new plane:

- The bottom of the interlock surface is not flat. We will create a plane and a planar surface and use it to trim the bottom.

- Click **Reference Geometry / Plane**, from the **Surfaces** tool tab.

- Select the **Front** plane for **First Reference**.

- Click the **Offset Distance** button and enter **.750in** for **Distance**.

- Place the new plane <u>below</u> the Front plane and click **OK**.

New plane

Interlock surface

12. Creating a new sketch:

- Select the new **plane1** and open a new sketch.

- Sketch a Rectangle around the model and add the dimensions shown to fully
position it.

Sketch rectangle

Dimension
to vertex

- While the sketch is still active click
the **Planar Surface** command from
the Surfaces tool tab (arrow).

- The rectangular sketch is converted
into a planar surface.

- Click **OK**.

13. Trimming the bottom of the ruled surface:

- Select the **Trim Surface** command from the Surfaces tool tab.

- Click the **Mutual** trim option (arrow). Select the **Planar Surface** and <u>all</u> of the surfaces along the perimeter of the Interlock Surface.

- Click the **Remove Selections** options and select the <u>inside</u> of the Interlock Surface plus all of its surfaces along the perimeter as indicated.

- Click **OK**.

Remove

Trim tools

Flat bottom

- The resulting Trimmed Surface. Change to the side view to verify the trim. The bottom of the interlock surface should be flat at this point.

14. Knitting the surfaces:

- Select the **Knit Surface** command from the Surfaces tool tab.

- Expand the FeatureManager tree and the Surfaces Bodies folders. Select the Parting Surface and all of the surfaces that belong to the Ruled Surface.

- Click **OK**.

Knit the Parting Surface with the Ruled Surfaces

15. Creating the tooling split sketch:

- Select the **Planar Surface** and open a new sketch.

Offset from Sketch1

- Locate the **Sketch1** under the Planar Surface and <u>Show</u> it.

- Create an offset of **.125in** (smaller) from the Sketch1.

.125

- **Exit** the sketch and click **Tooling Split**. The Tooling Split properties appears.

- For Block Size, enter the following:

       Upper block: 2.500in
       Lower block: 1.500in

- Click **OK**.

- There are 3 solid bodies on the FeatureManager tree: the Original part, the Upper Mold Block, and the Lower Mold Block.

- We will separate them in the next step.

16. Separating the solid bodies:

- Select **Insert / Features / Move-Copy**.

- Click the **Translate / Rotate** button at the bottom of the tree.

- Select the <u>bottom half</u> of the mold (the core) for Bodies to Move/Copy.

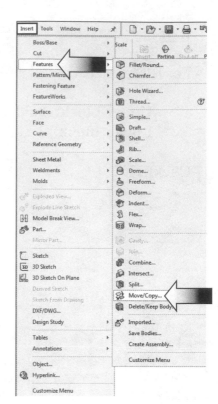

- Enter **-6.500in** for Delta Z distance. The core half is moved outward.

- Click **OK**.

- Repeat step number 16 and move the left half of the mold (the cavity).

- Move the cavity block about the **Delta Z** direction; use a distance of **8.250in**.

17. Making the body transparent:

- From the Feature tree, Expand the Solid Bodies folder.

- Right click on the Cavity body and select: **Change-Transparency** (arrow).

18. Hiding the surface bodies:

- The Core, Cavity, and other surfaces are still visible in the graphics making it difficult to see the molded part. We will need to hide them.

- From the FeatureManager tree, right click the Surface Bodies folder and select: **Hide**.

Hide all surfaces

19. Saving your work:

- Select **File / Save As**.

- Enter **Mold Manual Creation** for the file name.

- Click **Save**.

Creating Slides and Cores

Creating Slides and Cores
Mold Tooling Design

Many times undercuts cannot be avoided in plastic part designs. Features such as slots, triggers, locks and latches, etc. are often seen in plastic parts even though they may be more expensive to design and manufacture.

The Undercut Analysis tool can assist you with finding and visualizing trapped areas that may prevent the part from ejecting from the mold. These areas will need additional tooling such as lifters and cores to allow the part to be released from the tooling using the primary direction of pull.

A Slide or Core is a piece of tooling that slides out of the mold from the side, perpendicular to the direction that the part is ejected from the mold.

A sketch is used to define the shape and location of the core and it is normally drawn directly on the sides of the core or cavity block. Generally the sketch plane is parallel or perpendicular to the direction in which the side core travels away from the plastic part.

The Core command is used to extrude the sketch into a separate solid body and at the same time, it is subtracted from the core or cavity body. The Core command works like the Split command—they both divide a body into two or more bodies.

Additional tooling like Lifters, Core Pins, and Ejector Pins may be required to complete the mold design but this lesson will discuss the use of the standard mold tools to automate the creation of the Core and Cavity as well as the use of the Core command.

Creating Slides and Cores
Mold Tooling Design

<u>View Orientation Hot Keys</u>:

Ctrl + 1 = Front View
Ctrl + 2 = Back View
Ctrl + 3 = Left View
Ctrl + 4 = Right View
Ctrl + 5 = Top View
Ctrl + 6 = Bottom View
Ctrl + 7 = Isometric View
Ctrl + 8 = Normal To
      Selection

Dimensioning Standards: **ANSI**
Units: **INCHES** – 3 Decimals

Tools Needed:

	Scale		Parting Lines
	Parting Surfaces		Tooling Split
	Extruded Boss/Base		Core

1. Opening a part document:

- Select **File / Open**.

- Browse to the Training Files folder and open a part document named: **Slides and Cores.sldprt**

- One of the first things to do is to examine the model for potential problems that might prevent the core and cavity from separating.

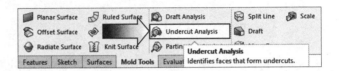

2. Analyzing the undercuts:

- The Undercut Analysis tool finds trapped areas in a model that cannot be ejected from the mold. These areas require a side core. When the main core and cavity are separated, the side core slides in a direction perpendicular to the motion of the main core and cavity, enabling the part to be ejected.

- Undercut Analysis works only on solid bodies, not surface bodies.

- Right click on one of the tool tabs and enable the **Mold Tools**.

- Select the **Undercut Analysis** command.

- For Direction of Pull, select the **Top** plane.

- Enable the **Highlight Occluded Regions** checkbox.

- Faces with different classifications are displayed in different colors in the graphics area. The faces are classified as follows:

Analysis Parameters explained:

* Direction1 Undercut:	Faces that are not visible from <u>above</u> the parting line.
* Direction2 Undercut:	Faces that are not visible from <u>below</u> the parting line.
* Occluded Undercut:	Faces that are not visible from above or below the parting line.
* Straddle Undercut:	Faces that have undercuts in both directions.
* No Undercut:	No Undercut.

---

* Direction of Pull:	(Not required if you select a Parting Line below.) Select a planar face, a linear edge, or an axis to define the draw direction, or select Coordinate input and set conditions along the X, Y, and Z axes.
* Parting Line:	Faces above the parting line are evaluated to determine if they are visible from above the parting line. Faces below the parting line are evaluated to determine if they are visible from below the parting line. This identifies depressions in the wall of the part that require a side core, and also helps you to identify sections of the parting line that you can modify to avoid the need for side cores.
* Adjustment Triad:	Manipulates the direction of pull to help you visualize ways to avoid or minimize problems with undercut regions. When you drag the rings of the triad in the graphics area, the direction of pull changes, face colors update dynamically, and the following read-only values appear in the Property-Manager:
	<ul><li>Angle with X axis</li><li>Angle with Y axis</li><li>Angle with Z axis</li></ul>
* Highlight Occluded Regions:	For faces that are only partially occluded, the analysis identifies those regions of the face that are occluded and those that are not. With this option cleared, the analysis identifies the entire face as being occluded.
* Undercut Faces:	Faces with different classifications are displayed in different colors in the graphics area. The results update in real time when you change the direction of pull.

- Click **Cancel** [X] to close the Undercut Analysis. A side core will be created to release the trapped areas in the mold.

3. Scaling the part:

- The Scale feature scales
  only the geometry of the model. It does not scale dimensions, sketches, or reference geometry. To temporarily restore the model to its un-scaled size, you can roll back or suppress the Scale feature.

- Use the Scale tool to account for the shrink factor when plastic cools.
  For odd shaped parts and glass filled plastics, you can specify nonlinear values.

- Click the **Scale** command on the Mold Tools toolbar.

- For Scale About, use the default **Centroid** option.

- Enable the **Uniform Scaling** checkbox.

- For Scale Factor, enter **1.02%** (arrow) (2% larger).

- Click **OK**.

- A Scale feature is like any other feature in the Feature-Manager design tree:
  it manipulates the geometry, but it does not change the definitions of features created before it was added.

4. Creating the parting lines:

- The parting lines lie along the edge of the molded part, between the core and the cavity surfaces. They are used to create the parting surfaces, and to separate the surfaces. The parting line is created after the model is scaled and proper draft is applied.

- Select the **Parting Lines** command from the Mold Tools toolbar (arrow).

- For Direction of Pull, select the **Top** plane (arrow).

- For Draft Angle, enter: **1.00deg**.

- Click the **Draft Analysis** button (arrow).

- The model includes a chain of edges that runs between positive and negative faces and the parting line segments are selected automatically; they are listed in Edges section.

- Click **OK**.

5. Creating the parting surfaces:

Parting Surfaces
Creates parting surfaces between core and cavity surfaces.

- The Parting surfaces split the mold cavity from the core. The parting lines and shut off surfaces before creating parting surfaces.

- The Shut-Off surfaces is used to shut-off the openings of the through holes in the plastic parts. Since this model does not have any through holes, we can skip this step and move on to creating the parting surfaces.

- Select the **Parting Surfaces** command (arrow).

- For Mold Parameters, select: **Perpendicular to Pull**.

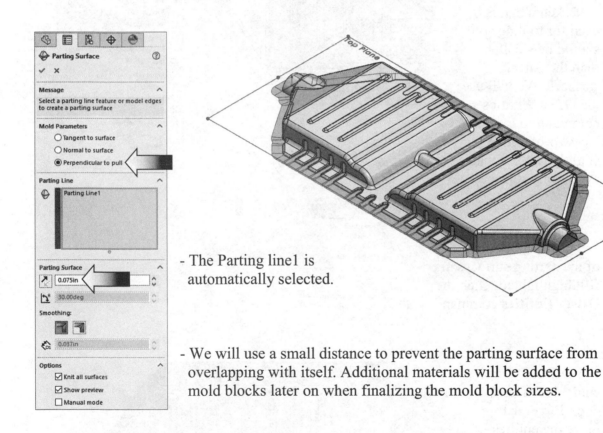

- The Parting line1 is automatically selected.

- We will use a small distance to prevent the parting surface from overlapping with itself. Additional materials will be added to the mold blocks later on when finalizing the mold block sizes.

- For Distance, enter **.075in.**; also enable the **Knit All Surfaces** checkbox.

- Click **OK**.

6. Creating a new sketch:

- Open a **new sketch** on the parting surfaces as noted. This sketch will be used in the following step to create the tooling split.

Sketch surface

- Right click on one of the outer edges of the parting surfaces and pick: **Select Open Loop** (arrow).

- The sketch that is used for tooling split should be smaller than the parting surfaces. We will use the Offset Entities command to create a new profile that is slightly smaller than the parting surfaces.

- Ensure that all outer edges of the parting surfaces are still highlighted; click the **Offset Entities** command.

- For Offset Distance, enter **.010in**. and click **Reverse** to place the entities on the <u>inside</u> of the parting surfaces.

.010

- Click **OK**.

7. Creating the tooling split:

- The Tooling Split tool uses the parting line, (shut off surfaces) and parting surfaces information to create the core and cavity, and allows the user to specify the block sizes.

- Exit the Sketch and select **Tooling Split** (arrow).

- For Upper Block size, enter: **.600in**

- For Lower Block size, enter: **.500in**

- Click **OK**.

8. Finalizing the upper block size:

- For clarity, right click the **Surface Bodies** folder and select **Hide**. Also right click the Parting Line1 and **Hide** it.

- Select the topmost surface of the upper block and open a new sketch.

Sketch face

- Sketch a **Corner Rectangle** as shown on the right.

- Add the **dimensions** as indicated to fully define the sketch.

- <u>Expand</u> the **Tooling Split1** on the FeatureManager tree, select the **Sketch1** and click **Convert Entities**.

- The Corner Rectangle and the Converted Entities formed a closed contour and will be used to add the additional material around the outside of the upper block.

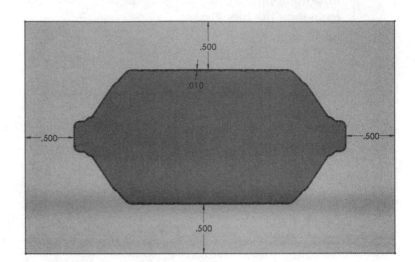

- Switch to the Features toolbar and select: **Extruded Boss-Base**.

- Use the default **Blind** type and a depth of **.600in**. Click **Reverse** direction.

- <u>Expand</u> the **Feature Scope** section and click the **Selected Bodies** option.

- Select the **upper block body** either in the graphics area or from the Solid Bodies folder. Enable the **Merge Result** box.

- Click **OK**.

9. Finalizing the lower block size:

- Switch to the **Isometric** view (Control+7).

- Hold the **Shift** key and press the **Up Arrow** twice to rotate the isometric view **180°** (the default angle is 90° when rotating with the Shift and an Arrow keys).

- Open a <u>new sketch</u> on the top surface of the lower block.

Sketch face

- Expand the Boss-Extrude1 and select the **Sketch2**.

- Click **Convert-Entities** to copy the entire sketch.

- Switch to the **Features** toolbar and select: **Extruded Boss-Base**.

- Use the default **Blind** type and a depth of **.500in**.

- Enable the **Reverse** button and the **Merge Result** checkbox.

- <u>Expand</u> the **Feature Scope** section and click the **Selected Bodies** option. Select the **lower block** body from the graphics area.

- Click **OK**.

10. Renaming the bodies and assigning materials:

- Expand the **Solid Bodies folder** and <u>rename</u> the solid bodies as indicated.

- Right click **each solid body** and assign the <u>material</u> as noted.

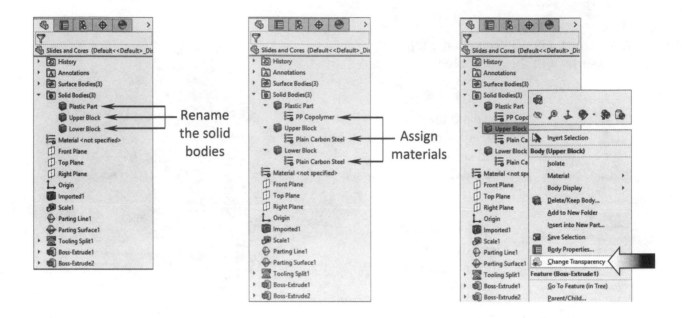

Rename the solid bodies

Assign materials

- Right click the <u>Upper Block</u> body and select: **Change Transparency** (arrow).

11. Creating the front slide core:

- We will create a couple of slide cores to capture the undercut features in the part.

- Open a <u>new sketch</u> on the **face** indicated.

- Sketch the profile shown. Use the mirror function to maintain the symmetric relations between the sketch entities.

Sketch face

- Notice the sketch is overlapped with the lower block by .073in? We will select only the upper block when making the core.

Mirror centerline

R.025

- <u>Exit the sketch</u> and select the **Core** command on the Mold Tools toolbar.

- The upper block (Boss-Extrude1) should be selected automatically.

- For Draft Angle, enter **1.00deg** and click the **Reverse** button.

- Use the default **Blind** type and enter **.730in** for depth.

- Verify the extrude direction is towards the part.

- Click **OK**.

12. Creating the back slide core:

- We will create an exploded view after both slide cores are created.

- Select the <u>side face</u> of the lower block and open a <u>new sketch</u>.

Sketch face

- Either convert, copy & paste the last sketch, or re-create it once again.

- Avoid snapping any endpoints of the model as it may cause the sketch to become over defined.

- Verify that the sketch is fully defined and positioned correctly.

- Exit the sketch and select the **Core** command
  on the Mold Tools toolbar.

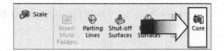

- The lower block (Boss-Extrude2) should
  be selected automatically.

- For Draft Angle,
  enter **1.00deg** and click
  the **Reverse** button.

- Use the default **Blind** type
  and enter **.730in** for depth.

- Verify the extrude direction is towards the part.

- Click **OK**.

13. Creating an exploded view:

- An exploded view in a
  multibody part shows
  the solid bodies spread
  out, but positioned to
  show how they fit
  together.

- You create exploded views
  by selecting the solid bodies and
  dragging the direction arrow in the
  graphics area, creating one or more explode steps.

- Select **Insert, Exploded View**.

Drag the **Y** arrowhead upward

- Select the upper block from the graphics area.

- Drag the **Y arrowhead** upward towards the vertical direction approx. 3.500 inches.

- Repeat the step above and explode the lower block and the 4 core blocks to the positions approximately as shown.

- Click **OK** when the exploded view is completed.

- Click the **Configuration-Manager** tab (arrow).

- Expand the default and double click **ExplView1** to Collapse it and right click to Edit.

14. Saving your work:

- Select **File, Save As**.

- Enter **Slides and Cores (Completed)** for the file name and press **Save**.

CHAPTER 20

Top-Down Assembly – Part 1

Top-Down Assembly – Part 1
Miniature Vise

This chapter will guide us through some techniques of creating new parts in the context of an assembly or Top Down mode.

By using the existing geometry of other parts such as their locations, features, and sizes to construct new components is referred to as In Context Assembly. This option greatly helps capture your design intent and reduces the time it takes to do a design change, having the parts update within themselves based on the way they were created.

While working in the top down assembly mode, every time a face or a plane is selected as a sketch plane to create a feature of the new part, the system automatically creates an INPLACE mate to reference the new part.

The Inplace mates can be suppressed so that components can be moved or repositioned and the Inplace mates can also be deleted as well; new mates can be added to establish new relationships with other components.

When a part is being edited in the Top Down Assembly mode, the Edit-Component icon is selected and the part's color changes to Blue (or Magenta depending on the color settings in the system options).

Upon the successful completion of this lesson, you will have a better understanding of the 2 assembly methods used in SOLIDWORKS: the traditional Bottom Up assembly (where parts are created separately, then inserted into an assembly document and mated together) and the dynamic Top-Down assembly (where parts can be created together in the context of an assembly).

Top-Down Assembly
Miniature Vise

Dimensioning Standards: **ANSI**

Units: **INCHES** – 3 Decimals

Tools Needed:

 Insert Sketch　　 Rectangle　　 Circle

 Dimension　　 Add Geometric Relations　　 Sketch Mirror

 Offset Entities　　 Planes　　 Fillet/Round

 Base/Boss Extrude　　 Loft　　 Edit Component

1. Starting a new assembly template:

- Select **File / New / Assembly**.

- Click **Cancel** 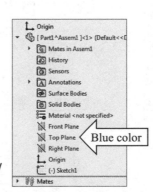 to exit the **Begin Assembly** mode.

- **Save** the new assembly document as **Mini Vise.sldasm**.

2. Creating the Base part:

- Select **Insert / Component / New Part**.

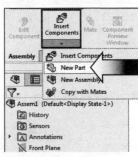

- Select the <u>Front</u> plane from the FeatureManager tree to reference the new part (Inplace1).

- A new component is created using a default name **[Part1^Assembly]<1>**.

- To rename the part, right click on the default name and select **Rename Part**.

- Enter: **Base** as the new name of the 1st part.

- The new part has the default Blue color.

- To change the part's color, go to: **Tools/ Options / System Options / Colors / Assembly / Edit Part.**

- A new sketch is created automatically when a new component is inserted.

- Sketch the profile shown below; keep the Origin at the lower right corner.

- Add the Dimensions or Relations needed to fully define the sketch.

3. Extruding the Base:

- Click or select **Insert / Boss-Base / Extrude**.

- Direction 1:   **Mid-Plane**.

- Extrude Depth: **.750 in.**

- Click **OK**.

4. Adding the side flanges:

- Select the <u>bottom face</u> of the base and open a new sketch .

- Sketch the profile below; use the Mirror option to keep the sketch entities symmetrical with the Centerline.

- Add dimensions as shown. (Hold the **Shift** key when adding the .625 dim.)

Sketch Face

5. Extruding the Flanges:

- Click or select **Insert / Boss-Base / Extrude**.

- Direction 1:   **Blind** (Reverse).

- Extrude Depth: **.200 in.**

- Click **OK**.

6. Adding the side cuts:

- Select the <u>face</u> as indicated and click or select **Insert / Sketch**.

- Sketch a Centerline ✎ starting at the Origin and click **Dynamic Mirror** 🔛.

- Sketch a rectangle and add the dimensions shown below.

Sketch Face

Coincident with corner

.100

.140

7. Extruding the side cuts:

- Click 🗐 or select **Insert / Cut / Extrude**.

- Direction 1: **Up-To-Surface**.

- Select the <u>face</u> as indicated.

- Click **OK**.

Select Face

.100

.080 .040

8. Creating an offset distance plane:

- Select the <u>face</u> as shown and click or select **Insert / Reference Geometry / Plane**.

- Click **Offset Distance** option.

- Enter **.150 in**. (the new plane is placed **away** from the face).

- Click **OK**.

Select Face

9. Creating the Fixed Jaw, sketch 1 of 4:

- Select the <u>new plane</u> and click or select **Insert / Sketch**.

- Sketch a rectangle approx. 2 inches above the origin.

<u>NOTE:</u> *The dimension .450 can be replaced with a centerline and a symmetric relation.*

More…

- Add dimensions to fully position the sketch.

- **Exit** the sketch ⬜ or select **Insert / Sketch**.

10. Creating the 2nd profile, sketch 2 of 4:

- Select the <u>face</u> as indicated and click ⬜ or select **Insert / Sketch**.

Sketch Face

- Hold the Control key and select the 4 edges as shown (or simply select the rectangular face and click the Convert Entities command).

- Click **Convert Entities** ⬛ on the Sketch-Tools toolbar.

Convert these 4 Edges

- The 4 selected edges are converted into a new 2D rectangle.

- **Exit** the sketch ⬛ or select **Insert / Sketch**.

11. Creating the 3D Guide Curves:

- Select **3D Sketch** from the Sketch tool tab 🔲 or select **Insert / 3D Sketch**.

- Sketch a 3-Point-Arc 🔲 approximately as shown and add the **Coincident** relation between the ends of the arc and the corners of the rectangles.

Tangent

A = 94.3° R = 3.679

- Add a **Perpendicular** relation between the <u>endpoint</u> of the arc and the <u>upper face</u> of the part as noted.

- Repeat the last step and create the other 3 arcs the same way.

Perpendicular
or Tangent

- **Exit** the 3D Sketch or select **Insert / 3D Sketch** (Control + Q).

12. Creating the Fixed Jaw loft:

- Click or select **Insert / Boss-Base / Loft**.

- Select the 2 sketch profiles as labeled (Profile 1 and Profile 2).
  (Click at or near the ends of the rectangles. SOLIDWORKS will select the nearest endpoints automatically.)

Loft Profile 1

Loft Profile 2

- If the preview shows a twisted transition, simply drag one of the connectors to the right corner will correct the problem.

More...

- Expand the **Guide Curve** section and select one of the guide curves in the 3D-Sketch.

- Because this sketch has multiple entities that are not connected with one another, you will have to click the OK button (the check mark) on the <u>SelectionManager</u> after selecting <u>each</u> arc (arrow).

- Click **OK**.

- The resulted loft with four guide curves.

13. Creating the Fixed Jaw Clamp:

- Select the <u>Front</u> plane from the FeatureManager tree and click [icon] or select **Insert / Sketch**.

- Sketch a Rectangle [icon] and add Dimensions [icon] and Relations [icon] as indicated. *(Add a horizontal centerline and a midpoint relation between the right endpoint and the model edge.)*

.375

Mid-Point with part

14. Extruding the Fixed Jaw Clamp:

- Click or select **Insert / Boss-Base / Extrude**.

- Direction 1: **Mid-Plane**.

- Extrude Depth: **1.250 in**.

- Click **OK**.

15. Creating the Lead Screw Hole:

- Select the <u>face</u> as indicated

and open a new sketch .

- Sketch a Circle and add the dimensions and relations as shown.

16. Extruding the Hole:

- Click or select **Insert / Cut / Extrude**.

- Direction 1: **Through All**.

- Click **OK**.

17. Adding Fillets:

- Click Fillet or select **Insert / Features /Fillets-Rounds**.

- Enter **.032 in**. for Radius.

- Select the edges as shown for Edges to Fillet.

- Click **OK**.

- The Base part is shown in Front and Back Isometric views.

18. Saving your work:

- Select **File / Save As / Base / Save**.

- Click to <u>exit</u> the Edit Component mode.

19. Creating a new component: The Slide Jaw

- Select **Insert / Component / New Part**.

- Rotate the model to a similar position as shown on the right. The planar surface on the left side will be used next.

- Select the <u>face</u> indicated as sketch plane for the new component (Inplace2). A new part and a new sketch are created in the FeatureManager tree.

- **Rename** the component to **Slide Jaw**.

Select face

- A new part is created in the FeatureManager tree and the sketch pencil is activated.

- An INPLACE mate is also created for the new component to reference its location.

20. Using the Offset Entities command:

- Select the **4 edges** of the model (as shown) and click **Offset Entities** [C] .

- Enter **.010 in**. for offset value. This offset distance between the 4 lines and the model edges will remain locked and get updated at the same time when the value is changed.

Edges to Offset*

.125

.010

.010

45.00°

(.010)

(.010)

.275

💡 **Offset Entities**

* The geometry of a model such as edges, faces, and other sketch entities can be offset or converted to use in the new part.
* The offset entities can be set to one direction or bidirectional.
* An On-Edge relation is created for each converted sketch entity.

- Sketch the rest of the profile and add the dimensions or relations needed to fully define the sketch.

Note:

The mirror option can *be used to help speed up the sketching process and keep the profile symmetrical at the same time.*

2 points (4X)
Horizontal

2 lines (2X)
Collinear

21. Extruding the Slide Jaw:

- Click on the Features toolbar or select:
  Insert / Base / Extrude.

- Direction 1: **Blind** and <u>reverse</u> direction.

- Extrude Depth: **1.000 in**.

- Click **OK**.

Transparency

* The transparent images are sometimes toggled on/off for clarity purposes.

22. Adding the support wall:

- Select the <u>face</u> indicated and open a new sketch or select **Insert / Sketch.**

Select Face

- Sketch the profile and add the dimensions and relation as shown below to fully define the sketch.

Ø.210

Vertical

Ø.210

.585   .225

.585

.225

NOTE: Remove the dimension .585 if the Concentric relation is used.

23. Extruding the Support Wall:

- Click on the Features toolbar or select **Insert / Base / Extrude.**

- Direction 1: **Blind** and <u>reverse</u> direction.

- Extrude Depth: **.375 in**.

- Click **OK**.

- The Support wall is built with a guide hole.

24. Creating a new work plane:

- Select the face as indicated and click or select:
  Insert / Reference Geometry / Plane.

- Enter **.150 in**. for Offset Distance and place the new plane on the **outside**.

Select Face

New Plane

25. Creating the Slide Jaw 1ˢᵗ sketch:

- Open a new sketch on the <u>new plane</u> or select **Insert / Sketch.**

- Sketch a rectangle and add the dimensions as shown to fully define the sketch.

- **Exit** the sketch or select: **Insert / Sketch.**

26. Creating the Slide Jaw 2ⁿᵈ sketch:

- Select the <u>face</u> indicated and open a new sketch 🔲 or select: **Insert / Sketch.**

- Sketch a rectangle 🔲 as shown.

Sketch Face

Mid-Point relation between endpoint and edge

.750

.500

.750

.500

- Add dimensions 🔲 to fully define the sketch.

- **Exit** the Sketch 🔲 or select **Insert / Sketch.**

27. Creating the Guide Curve to connect the two sketches:

- Select the <u>Right</u> plane of the part from the FeatureManager tree.

- Click 🔲 to open a new sketch or select: **Insert / Sketch.**

💡 **Guide Curves**

* Guide curves are used to control the profile from twisting as the sketch is swept along the path.
* Guide curves are also used in Sweep to shape the 3D Features.
* Each profile is Pierced or coincident with the guide curve.

- Sketch either a **Centerpoint Arc** or a **3-Point Arc** that connects the two sketches.

- Add the Relations as shown below to fully define the sketch.

3 Point Arc or Centerpoint Arc

Pierce Relation

Pierce Relation

Add centerline and make it tangent w/arc.

- **Exit** the sketch or Select **Insert / !**

28. Creating the Slide Jaw Loft:

- Click on the Features toolbar or select **Insert / Boss / Loft.**

- Select the upper corners of the 2 rectangular sketches to use as Loft Profiles.

- Expand the **Start/End Constraints** section and set the following:

Select line for Direction Vector.

* Start Constraint: **Normal to Profile**

* End Constraint: **Direction Vector** and select the centerline as noted

- Expand the Guide Curves dialog box and select the Arc (sketch4) to use as a Guide Curve.

(See next page for details.)

Start/End Constraints

* The Start constraint and End constraint option applies a constraint to control tangency to the start and end profiles.

* The Direction Vector option applies a tangency constraint based on a selected entity used as a direction vector.

OPTIONAL:
4 Guide Curves
(in one 3D Sketch)

Tangent

Start/End Constraints

Start constraint:

Normal To Profile

1

End constraint:

Direction Vector

Line2@Sketch4@Slide Jaw-1@M

1

- Click **OK**.

29. Creating the Clamp block:

- Select the part's <u>Right</u> plane from the FeatureManager tree.

- Click [ ] to open a new sketch or select: **Insert / Sketch.**

- Sketch a Rectangle and add Dimensions as shown.

- Add a Centerline in the middle of the rectangle and position it on the Mid-Point of the vertical edge.

Mid-Point with vertical edge

30. Extruding the Clamp block:

- Click or select **Insert / Boss-Base Extrude.**

- Direction 1: **Mid-Plane**.

- Extrude Depth: **1.250 in**.

- Merge Result: **Enabled**.

- Click **OK**.

31. Which option is better?

- Instead of using the Mid-Plane extrude, the **Up-To-Surface** option can be used to link the length dimensions of the 2 Clamp Blocks together.

- Right click on the last Extruded feature and select **Edit Feature**.

- Change **Direction 1** from Mid-Plane to **Up-To-Surface** and select the face on the left side.

- Change **Direction 2** to **Up-To-Surface** and select the face on the right side as indicated.

- Click **OK**.

 Up-To-Surface

* Extends the feature from the sketch plane to the selected surface.
* When the driving surface is changed in length, the referenced extruded feature will also be reflected.

Direction 2:
Up-To-Surface
(Right side)

Direction 1:
Up-To-Surface
(Left side)

32. Adding fillets:

- Click or select **Insert / Features / Fillet-Round.**

- Enter **.032** for Radius value.

- Select the edges as shown.

- Click **OK**.

- The fillets are shown in the Front and Back Isometric views for clarity.

33. Creating the internal threads:

- Starting with the sweep path.

- Select the <u>face</u> indicated and open a new sketch or select: **Insert / Sketch.**

Sketch face ———

- Sketch a Circle ⊡ that is Concentric with the hole. (Converting the ID of the hole is another good way to link the diameter of the circle to the hole's diameter.)

- Add a **Ø.210** dimension ✎ to fully define.

Ø.210 ———

🔆 **Wake up Center Points**

* With the Circle tool selected, hover the mouse cursor over the circumference of the hole; the 4 quadrant points appear, and the center-point of the circle is visible for snapping.

- Select **Insert / Curve / Helix-Spiral.**

- Enter the following parameters:

Defined by:	**Pitch and Revolution**.
Pitch:	**.080 in**.
Revolution:	**5.000**.
Starting Angle:	**90.00 deg**.
Reverse Direction:	**Enabled**.
Clockwise:	**Selected**.

- Click **OK**.

Click **View / Hide-Show / Curves** if the helix is not visible.

- Sketching the Sweep Profile:

- Select the part's <u>Right</u> plane from the Feature tree and open a new sketch or select: **Insert / Sketch.**

- Sketch the profile as shown below.
  (Use Mirror to keep the entities symmetrical.)

- Add Dimensions and Relations to fully define the sketch. Pierce the endpoint of the centerline to the 1st revolution of the helix.

- Change to the Front view and Hidden Lines Visible option .

- **<u>Exit</u>** the sketch or Select **Insert / Sketch.**

34. Sweeping the thread Profile along the Helix:

- Click or select **Insert / Cut / Sweep.**

- For Sweep Profile, select the **Thread Profile**.

- For Sweep Path, select the **Helix**.

- Click **OK**.

35. Creating a Section View:

- Click the **Section View** command or select **View / Display / Section View**.

- Select the **Front** plane of the assembly for cutting plane.

- Verify the details of the threads.

- Click the **Section View** icon again to turn it off.

36. Saving your work:

- Save your work once again using the same file name: **Mini-Vise.sldasm**

- Overwrite the old file when prompted.

37. Assembly Exploded view (Optional):

- Create the additional components: Lead Screw, Crank Handle, and Crank Knob using the Top Down Assembly method.

- Create an assembly exploded view as shown (details on next page).

** To define the Helix, use the
Height and Pitch option.*
Height = 2.116
Pitch   = .080

LEAD SCREW

CRANK HANDLE

CRANK KNOB

Questions for Review

Top-Down Assembly

1. New parts can be created in context of an assembly.
    a. True
    b. False

2. Geometry of other components such as model edges, hole diameters, and locations, etc. can be used to construct a new part.
    a. True
    b. False

3. Part documents can be inserted into an assembly using:
    a. Insert menu
    b. Windows Explorer
    c. Drag and drop from an open window
    d. All of the above

4. The suffix (f) next to the first part's name in the FeatureManager tree stands for:
    a. Fail
    b. Fixed
    c. Float

5. When inserting new components into an assembly, the Inplace mates are created by the user.
    a. True
    b. False

6. Either in the part or assembly mode, the guide curves are used to help control the profiles from twisting, as they are swept along the path.
    a. True
    b. False

7. Centerpoint Arcs are drawn from its center, then radius, and angle.
    a. True
    b. False

8. The Link Values option allows a user to link only two dimensions at a time.
    a. True
    b. False

7. TRUE
8. FALSE
5. FALSE
6. TRUE
3. D
4. B
1. TRUE
2. TRUE

CHAPTER 21

Top-Down Assembly – Part 2

Top-Down Assembly – Part 2
Water Control Valve

When a component is built in the context of an assembly external references are created to reference how it was constructed, and which plane or surface was used to create it with. Starting from the very <u>Top</u> level assembly, information regarding the new component is added and flows <u>Down</u> to the component level and gets repeated every time a new component is added.

For example: The mounting holes in the second part can be converted from the first, so that the hole diameters and the location dimensions are the same for both parts. When the holes in the 1st part are changed, the holes in the 2nd part would also change. Thus, the sketch of the holes in the 2nd part is defined in the assembly, not by sketching and dimensioning them as in the part mode.

Using the Top Down assembly design, one of the better approaches is to use the geometry of the existing parts to create the new. This way several parts can be controlled and changed at the same time.

There are many advantages for creating parts in Top Down mode, and just to mention a few: not only is this method much quicker than the others due to the ability to use existing geometry to reference the new parts, but because all the parts are always visible in the assembly to help develop the <u>Form</u> of the new part and how it is supposed to <u>Fit</u> with other parts, it is more predictable how it is going to <u>Function</u>. Interference, friction and or clearance fits can be created and controlled within the very same screen.

However, there are a few things to consider when designing in Top Down mode:
* External references are created to the geometry that the new part is referenced to, and that means:
  * When changes occur, the assembly updates all of its internal parts, and if drawings were made from these parts earlier, they will get updated as well.

Top Down Assembly
Water Control Valve

View Orientation Hot Keys:

Ctrl + 1 = Front View
Ctrl + 2 = Back View
Ctrl + 3 = Left View
Ctrl + 4 = Right View
Ctrl + 5 = Top View
Ctrl + 6 = Bottom View
Ctrl + 7 = Isometric View
Ctrl + 8 = Normal To
        Selection

Dimensioning Standards: **ANSI**

Units: **INCHES** – 3 Decimals

Tools Needed:

Insert Sketch	Line	Circle
Add Geometric Relations	Sketch Fillet	Trim
Dimension	Centerline	Fillet/Round
Base/Boss Revolve	Extruded Boss/Base	Edit Component

1. Starting with a new Assembly Template:

- Click **File / New**.

- Select an **Assembly** template either from the Template or the Tutorial tab.

- Click the **Advance** button at the lower left corner of this dialog box if you do not see the similar templates.

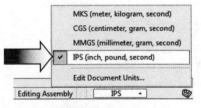

- The **Begin Assembly** dialog appears on the left side; click **Cancel**. We're going to use a different approach to create the new components.

- At the bottom right of the screen, set the Units to **IPS (Inch, Pound, Second).**

- From the **Assembly tab**, click the **drop arrow** below the Insert Components command and select: **New Part**.

- Creating components in context of an assembly will require a few additional steps:

  a/. A new part is inserted into an Assembly and the Part's name is entered.

  b/. A plane is selected at this time to reference the new part.

  c/. The Edit Component command is activated and the Sketch mode is enabled for the plane selected in step b.

  d/. The active part will change to the blue color by default.

2. Creating the 1st component:

- When the symbol appears next to your mouse cursor, select the **Front** plane on the Feature tree. An Inplace mate is created to reference the new part.

- Press **Control + 1** to switch to the Front orientation.

NOTE: *To automatically rotate normal to the sketch plane, go to System* **Options / Sketch**, *and enable the checkbox:* **Auto Rotate View Normal to...**

3. Sketching the Base Profile:

- Sketch the profile <u>above</u> the origin.

- Add the dimensions shown. The diameter dimensions are shown as Virtual Diameters.

- Click **Revolve** 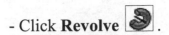.

- The centerline should be selected automatically.

- Use the default Blind option and revolve the sketch one complete revolution.

- Click **OK**.

4. Adding the Inlet Flange:

- Select the <u>face</u> indicated and open a new sketch.

— Sketch face

- Sketch the profile shown below. Use the Dynamic Mirror option to help speed up the sketching process.

- Add the dimensions and relations needed to fully define the sketch.

- The number of places are added to help clarify the sketch; you do not have to add them.

- Click **Extruded Boss-Base**.

- Use the **Blind** extrude option.

- Enter **.650"** for thickness.

- Enable the **Merge Result** checkbox.

- Click **OK**.

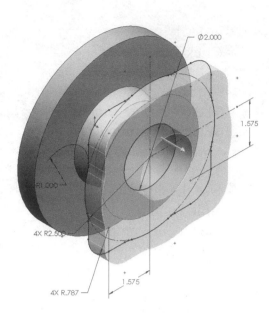

5. Adding the mounting holes:

- Select the <u>face</u> indicated and open a new sketch.

- Sketch a couple of center-lines to help locate the center and directions for this sketch.

- Add a circle and either mirror it or circular pattern it 4 times around.

Sketch face

Coincident

- Add the dimensions and relations needed to fully define the sketch.

- Click **Extruded Cut**.

- Select the **Up-To-Next** extrude option.

- Click **OK**.

6. Adding other mounting holes:

- Select the <u>face</u> as noted and open a new sketch.

- Sketch a circle and **convert** it to **construction** (click the For-Construction checkbox on the Feature tree).

- Add a couple of **centerlines** as shown.

- Sketch a **smaller circle** that's coincident with the construction circle and the endpoint of the centerline.

- Use the **Circular-Sketch-Pattern** option to array the small circle 4 times around.

- Add the dimensions and relations needed to fully define this sketch.

NOTE: *An additional Vertical or Horizontal relation between the centers of the small circles is needed when using the 2D sketch pattern.*

The center of the circle is coincident with the construction circle and the endpoint of the angled centerline.

Sketch face

45.00°

Ø4.750

Ø.525

Coincident

Horizontal

- Click **Extruded Cut**.

- Use the **Up-To-Next** extrude option to ensure that the cut only goes through the thickness of the flange.

- Click **OK**.

- Rotate the view to verify the cut result.

7. Adding the .032" chamfers:

- Click **Chamfer** (below the Fillet command).

- Enter **.032"** for Depth and use the default **45°** angle.

- Select the edges of the 8 holes and the 2 edges of the round flange.

- To un-select an edge simply click it once again.

- Click **OK**.

Fillet edges

8. Adding the .060" fillets:

- Click the **Fillet** command.

- Enter **.060"** for radius size.

Fillet edges

- Select the **3 edges** of the transition body.

- Enable the Full Preview checkbox.

- Change to the Top orientation (Control + 5) to verify the selection.

- Click **OK**.

9. Adding the .125" chamfers:

- Click **Chamfer** once again.

- Enter **.125"** for Depth and use the default **45°** angle.

- Select the **2 edges** of the center hole.

- Selecting the face of the hole would get the same result as selecting its 2 edges.

- Click **OK**.

Fillet edges

10. Exiting the Edit Component mode:

- Click off the **Edit Component** button to return to the Edit Assembly mode.

- When the Edit Component command is <u>not active</u>, the part's color changes back to its default color (gray).

11. Renaming the component:

- Right click the name of the part (Part1) from the FeatureManager and select: **Rename Part** (arrow).

- Enter **Outlet Housing** and press enter.

12. Saving as Virtual Component:

- Virtual components are quite useful in the Top-Down Assembly mode. These components are saved internally, or embedded in the assembly document, instead of as separate part or sub-assembly documents.

- Click **File /Save As**.

- Enter **Water Control Valve** for the file name and press **Save**.

- The Save As Virtual Component dialog appears; click the **Save Internally** option (Inside the Assembly) and click **OK**.

- When the parent assembly is opened, all virtual components are also loaded into RAM. The virtual components can then be opened so that the detail drawings can be generated from them, or they can simply be saved as external part documents to share with others.

13. Creating the 2nd component:

- Click the **New Part** command under the Insert Components drop down.

- When the symbol appears next to your mouse cursor, click the <u>Face</u> of the flange as indicated.

- At this point, another Inplace mate is created for the new part.

- A new (Part2) component is created on the FeatureManager tree.

Select face

- The Outlet Housing changes to transparent (inactive).

- The **Edit Component** command is activated.

- A new Sketch is also enabled automatically.

- Right click on one of the outer edges and pick **Select Tangency**.

Convert Entities
Converts selected model edges or
sketch entities into sketch segments.

Right click and pick:
Select Tangency

- Press **Convert Entities**.
  The selected edges are converted
  to new sketch entities.

*Convert Entities
(inside diameters)*

- Convert also the **circular edges** of the 4 holes.

- Click **Extruded Boss-Base**.

- Use the default **Blind** extrude option.

- Enter **.650"** for thickness.

- Click **OK**.

- The new flange is created by converting the geometry of the 1st component.

- If the 1st component is changed the 2nd component will also get updated. We will take a look at some of the changes toward the end of this chapter.

14. Creating the transition body:

- Select the part's <u>Right</u> plane and open a new sketch.

- Sketch the profile shown below and add the dimensions and any relations needed to fully define this sketch. (Add the R2.00 fillet after fully defined.)

- Click **Revolve Boss-Base**.

- The revolve centerline is selected by default.

- Use the default settings:

   * **Blind**
   * **360deg**

- Click **OK**.

15. Adding another mounting flange:

- Select the <u>face</u> as indicated and open a new sketch.

- Hold the control key and select the circular edge of the round flange <u>and</u> the 4 edges of its mounting holes (arrow).

Sketch face

- Click **Convert Entities**.

Convert 5 edges
(4 inside diameters
and 1 outer flange)

- The selected edges are converted and projected onto the sketch face.

- Each sketch entity is linked to the original geometry where it was converted from. A relation called On-Edge is added to maintain their parent and child relationships.

- Click **Extruded Boss-Base**.

- Use the default **Blind** option.

- Enter **.650"** for depth.

- Extrude direction is outward.

- Click **OK**.

16. Adding an Offset-Distance plane:

- Select the <u>Top</u> plane of the part from the FeatureManager tree.

- From the Features tab, click **Reference Geometry / Plane**.

- The Offset Distance option should be selected by default; enter **1.625"** for distance, and place the new plane <u>above</u> the Top plane.

- Click **OK**.

17. Adding a circular boss:

- Select the new <u>Plane1</u> and open a new sketch.

- Sketch a **circle** approx. as shown.

- Add the **1.50"** diameter dimension and the **1.00"** locating dimension.

- Add the **horizontal** relation as noted to fully define the sketch.

Horizontal

- Click **Extruded Boss-Base**.

- Use the default **Blind** option.

- Enter: **.875"** for depth.

- Click **OK**.

18. Adding a thermostat valve mount:

- Open a new sketch on the <u>upper face</u> of the boss.

- Sketch the shape of the thermostat shown below.

- Use the mirror function to help maintain the symmetrical relationships of the sketch entities.

- Add the dimensions and the relations as noted in the drawing below.

Sketch face

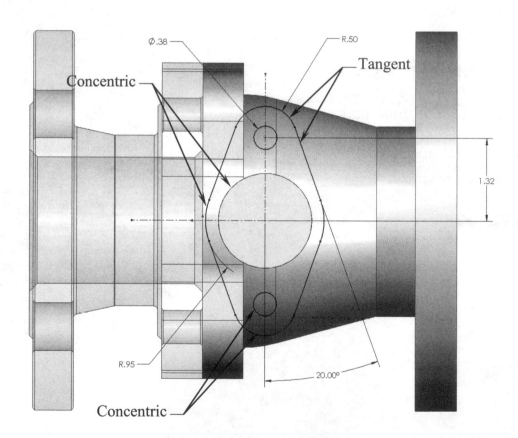

- The centers of the circles should be vertical with each other.

- If the mirror feature was not used, then be sure to add the symmetric relations to fully define this sketch.

- Click **Extruded Boss-Base**.

- For **Direction 1**: Use **Blind** and the depth of **.250"**.

Direction 2:
Select this Surface

- For **Direction 2**: Use **Up-To-Surface** option and select the planar face as indicated.

- Click **OK**.

19. Adding a hole:

- Open a new sketch on the upper face of the thermostat.

- Sketch a **Ø1.00 circle** and add a **concentric** relation to center it.

- Make use of the "Wake-up the Entities Snap Mode."

(With the Circle command selected, hover the mouse cursor over the circular edge of the radius. The appropriate snap-entities will appear to allow snapping to the existing geometry.)

Sketch face

- Click **Extruded Cut**.

- Click the **Reverse** direction button.

- Select the option **Up-To-Next** from the list.

- Click **OK**.

20. Adding the .080" fillets:

Select 6 edges

- From the Features tab, click **Fillet**.

- Enter **.080"** for radius size.

- Select the **6 edges** as indicated.

- The Tangent Propagation checkbox should be selected.

- Click **OK**.

21. Adding the .032" chamfers:

- Click **Chamfer** under the Fillet command.

- Enter **.032"** for depth.

- Use the default **45deg** angle.

- Select the **edges** of the holes as shown in the preview image. (23 edges total.)

- The same result can be achieved by selecting the inner faces of the holes.

- Click **OK**.

Only select the upper edge, not the bottom

Select edges front & back

22. Adding the .125" chamfers:

- Click **Chamfer** once again.

- Enter **.125"** for depth.

- Use the same **45deg** angle.

- Select the **2 edges** of the center hole. (Selecting the face of the hole would get the same result.)

- Click **OK**.

Select 2 edges

23. Exiting the Edit Component mode:

- On the Assembly toolbar, click off the **Edit Component** command (arrow).

24. Applying dimension changes:

- Expand the component Outlet Housing and double click on the feature **Revolve1**.

- Change the flange diameter **from 6.00" to 7.00"**

- Click the **Rebuild** (the green traffic light) to execute the change*.

- Notice the dimension change also updates the flange diameter on the right.

Changing this diameter will also change the other diameter

* Press <u>undo</u> to switch the dimension back to its original value, or double click the same dimension, re-enter the previous value, and click Rebuild again.

- Double click the feature **Cut-Extrude1** of the same part.

Double click inside the hole to see its dimensions

- Doubling the inner face of the small hole would also display its dimensions.

- Locate the dimension Ø**.750**.

- Change the hole diameter **from .750" to .500"**.

- Click the **Rebuild** green traffic light to execute the change*.

Double click the
Ø.750" dimension
and change to .500"

Changing this diameter will also
change the other diameters

- Notice the dimension change also updates the hole diameters on the right.

    * Press <u>undo</u> to switch the dimension back to its
      Original value, or double click the same dimension,
      re-enter the previous value, and click Rebuild again.

25. Viewing the External Reference Symbols:

- Expand the 2nd part, the **Inlet Housing**.

- Some of the features have the External reference
  symbols next to their names. These references were
  created automatically when we converted the entities
  of the Part1 to create the new sketch for the Part2.
  They are called On-Edge relations.

- An external reference is also created when we add
  a dimension or a relation between Part1 and Part2.

26. Inserting other components:

- Due to the length of this lesson, we are going to insert and mate the rest of the components that belong to this assembly.

- Browse to the Training Files folder and insert the components as labeled below.

- Create the mates that are appropriate for each component. The non-moving components will get 3 mates, and the moving components will need only two.

27. Optional:

a/. Create a section view to verify how the components were mated.

b/. To center the 2 components, it is best to use the Width mate option.

c/. Change / correct any mates or geometry that would cause the interferences.

- Add the Explode-Line Sketch as shown.

- When adding the explode lines, pay attention to the direction arrows, flip or reverse them before completing each line.

28. Saving your work:

- Click **File / Save As**.

- Enter **Water Control Valve.sldasm** for the name of the file.

- Press **Save**.

Questions for Review

Top-Down Assembly

1. In Top Down mode, when a plane is selected to sketch the new part, SOLIDWORKS will create an Inplace mate to reference the new part.
   a. True
   b. False

2. After the plane is selected, SOLIDWORKS will also activate the Edit Component command and the Sketch mode at the same time.
   a. True
   b. False

3. The option Auto-Rotate Normal to the Sketch is not available to set as the default.
   a. Truc
   b. False

4. The Convert Entities command can only be used when the Sketch pencil is turned off.
   a. True
   b. False

5. The Virtual part is saved / cmbedded inside an assembly document.
   a. True
   b. False

6. When editing a part in Top Down mode, both the current part and other parts can be filleted at the same time.
   a. True
   b. False

7. The Edit Component command should be left active prior to inserting a new part.
   a. True
   b. False

8. When a dimension is changed, any reference geometry of other parts should also change.
   a. True
   b. False

7. FALSE	8. TRUE
5. TRUE	6. FALSE
3. FALSE	4. FALSE
1. TRUE	2. TRUE

CHAPTER 22

External References & Repair Errors

External References & Repair Errors

An **external reference** is created when one document is dependent on another document for its solution. If the referenced document changes, the dependent document changes also.

In an assembly, you can create an **in-context** feature on one component that references the geometry of another component. This in-context feature has an external reference to the other component. If you change the geometry on the referenced component, the associated in-context feature changes accordingly.

The External Symbols:

->	.... External Reference	**?**	.... Out Of Context
(+)	.... Over Defined	*****	.... Reference Locked
X	.... Reference Brokcn		

-> External Reference:
The part itself or some of its entities are depending on the geometry of other parts for their solutions.

? Out Of Context:
The part or its features are not solved, not up-to-date or disconnected from its assembly.

(+) Over Defined:
The Dimensions or Relations of the sketch are conflicting; redundant dimensions or wrong relations were used.

* Reference Locked:
Lock the external references on a part; the existing references no longer update and the part will not accept any new references from that point.

X Reference Broken:
The references between the part and the others are broken. Changes done to the part will not affect the others.

External References & Repair Errors

Dimensioning Standards: **ANSI**
Units: **INCHES** – 3 Decimals

View Orientation Hot Keys:

Ctrl + 1 = Front View
Ctrl + 2 = Back View
Ctrl + 3 = Left View
Ctrl + 4 = Right View
Ctrl + 5 = Top View
Ctrl + 6 = Bottom View
Ctrl + 7 = Isometric View
Ctrl + 8 = Normal To
           Selection

External Reference Symbols:

->	External Reference	?	Out of Context
(+	Over Defined	*	External Reference Locked
X	External Reference Broken	⌁	Display/Delete Relations

Understanding & Removing External References

1. Opening an existing part: Double Joint

- Go to: Training Files folder

- Open the part named: **Double Joint**.

- The **What's Wrong** dialog appears
  displaying the current errors. Close it.

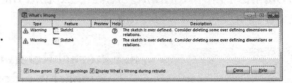

2. Listing the External References:

- Right click on the part's name and select **List External Refs**.

- Enable the checkbox:
  List External References.

Several Lines and Arcs were converted from other
parts in an assembly; deleting these references will
cause those entities to become Dangling.

3. Removing External References:

- Click **Break All** to remove the
  external references.

- Click **OK** [OK] .

4. Understanding the External Symbols:

-> .... External Reference
? .... Out Of Context
(+) .... Over Defined
* .... Reference Locked
X .... Reference Broken

(From the FeatureManager tree, expand the first two features to see their sketches.)

💡 **Error Colors**

- **Olive Green:** Dangling
  (Missing references/detached).

- **Red:** Over Defined
  (Wrong Relations/Dimension).

- **Yellow:** Not Solved
  (Relation is conflicting with dimensions).

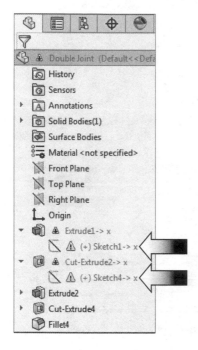

a. External Reference ->:
The model itself or some of its entities are depending on the geometry of other parts for their solutions.

b. Out Of Context ?:
The model or its features are not solved, not up-to-date, or disconnected from its assembly.

c. Over Defined (+):
The Dimensions or Relations of the sketch are conflicting; Redundant dimensions or wrong relations were used.

d. Reference Locked *️
Lock the external references on a model, the existing references no longer update - and - the model will accept any new references from that point.

e. Reference Broken X:
The references between the part and the others are broken. Changes done to the part will not affect the others.

5. Viewing the existing Relations:

- Right-click on the 1st sketch and select **Edit-Sketch** ⬛.

- Click 🔧 or select **Tools / Relations / Display-Delete.**

- The Ø.325 is shown in Olive Green color; this indicates either a relation is wrong or the sketch is over dimensioned.

- Change the Relations Filter to **All In This Sketch.**

6. Repairing the 1st sketch:

- Click **Display / Delete Relations** and delete the **Coradial** relation (the Circle is Coradial with an entity that no longer exists).

- Delete all relations that have the External Relations (->X) next to their names.

Delete Coradial
(Wrong Relation)

Delete External
Relations (-> X)

- Add the new Dimension as noted to fully define the sketch.

Add new dimensions to fully define the Sketch.

.537

.676

Ø.325

The sketch is no longer over defined.

| 0.65in | 0.112in | 0in | Fully Defined |

- A message on the lower right indicates the sketch is no longer over defined.

- Click **OK** and **Exit** the sketch. There are still some other errors in the part.

7. Repairing the 2nd sketch:

- Right click on the 2nd sketch and select **Edit-Sketch** 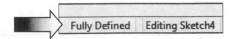.

- Select the **Display / Delete Relations** command once again.

- Delete the External Dimensions and Relations that were created in context of other parts.

Delete Collinear 15 (External Relation)

Delete External Dimension

Delete the Mirror point

- Delete also the two "Mirror-Points" as noted.

- **Exit** the sketch when the message:
Fully Defined is displayed on the lower right.

Fully Defined | Editing Sketch4

8. Rebuilding the model:

- Press **Rebuild** 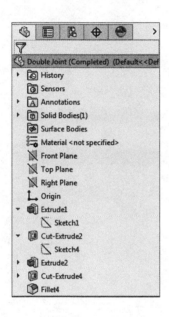 to re-generate the model.

- Verify that the part has no rebuild errors and there should not be any external reference symbols in the FeatureManager tree.

9. Saving your work:

- Select **File / Save As.**

- Enter **Breaking External References** for the file name.

- Click **Save**.

Questions for Review

External References

1. The symbol **->** next to a file name means:
   a. Dangling dimension
   b. External reference
   c. Not solved

2. The symbol **?** next to a file name means:
   a. The part cannot be found
   b. Wrong mates
   c. Wrong relations
   d. Out of context

3. The symbol **X** next to a file or a feature name means:
   a. The part or feature is wrong
   b. The part or feature is deleted
   c. The external references are broken

4. The symbol ***** next to a file name means:
   a. External references are locked
   b. Select all references
   c. Deselect all references

5. The symbol ***X** next to a feature name means:
   a. The feature is fully defined
   b. The feature is over defined
   c. The feature is under defined
   d. None of the above

6. The Olive-Green color in a sketch means:
   a. The sketch entity is selected
   b. The sketch entity is being copied
   c. The sketch has dangling entities, relations or dimensions

7. The dangling dimension can be "re-attached" simply by dragging its handle point to a sketch line or a model edge.
   a. True
   b. False

7. TRUE
5. D
6. C
3. C
4. A
1. B
2. D

Understanding and Repairing Part Errors

When an error occurs, SOLIDWORKS will try and solve it based on the settings below:

To pre-set the rebuild action:

A. Click **Options** 🔘 or **Tools / Options / General**.

B. Select **Stop, Continue,** or **Prompt** for **When rebuild error occurs**, and then click **OK**. With **Stop** or **Prompt**. The rebuild action stops for each error so you can fix feature failures one at a time.

Indicates an error with the model. This icon appears on the document name at the top of the FeatureManager design tree, and on the feature that contains the error. The text of the part or feature is displayed in **red** color.

Indicates an error with a feature. This icon appears on the feature name in the FeatureManager design tree. The text of the feature is displayed in **red** color.

Indicates a warning underneath the node indicated. This icon appears on the document name at the top of the FeatureManager design tree and on the parent feature in the FeatureManager design tree whose child feature issued the error. The text of the feature is displayed in **olive green** color.

Indicates a warning with a **feature** or **sketch**. This icon appears on the specific feature in the FeatureManager design tree that issued the warning. The text of the feature or sketch is displayed in **olive green** color.

1. Opening a part document:

- Insert the Training File folder, browse to the Repair Errors folder, and open the part named: **Repair Errors**.

- When opening a document that contains errors, the What's Wrong dialog box will appear and display where the errors are located and suggest some solutions in solving them.

- **Expand** each feature on the FeatureManager tree and hover the pointer over the **Sketch3**.

- A description about the error or the warning is displayed in the tooltip. This is the same as right clicking on the error and selecting the What's Wrong option.

2. Repairing the 1st error:

- Select the **Sketch3** and click **Edit Sketch**.

- Click the **Display/Delete Relation** command .

- Change the Filter to **All In This Sketch** (arrow).

- There are some dangling coincident relations in this sketch; they appear in the olive green color.

- Select the **Coincident3**. One endpoint of a line is coincident to a missing edge. **Delete** this Coincident3.

One endpoint of a line is coincident to a missing edge

- **Delete** also the **Coincident5**. It's missing the same edge as the previous relation.

- The dimension **6.500** is also dangling. It was measured to a missing entity; **delete it**.

- A message on the bottom right of the screen appears, indicating that: **The sketch can now find a valid solution**.

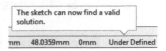

Delete this
Dangling dimension —

Add 3 new —
dimensions

- **Add** 3 new dimensions as indicated to fully define this sketch. **Exit** the sketch when completed.

- After exiting the sketch, SOLIDWORKS continues to report other errors still remaining in the part. The **Sketch4** and **Sketch11** still need to be repaired.

- The What's Wrong dialog box pops up displaying the same description about the selected error. Enable the **Show Warnings** checkbox (arrow) to see the warnings after each rebuild.

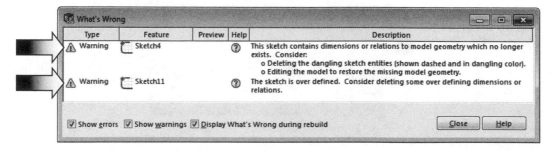

- **Close** the What's Wrong dialog box.

3. Repairing the 2nd error:

- Click the **Sketch4** (under the Sketched Bend1 feature) and select: **Edit Sketch**.

- Select the **Display / Delete Relations** command once again.

- The **Coincident1** is dangling and has the Olive Green color. **Delete it**.

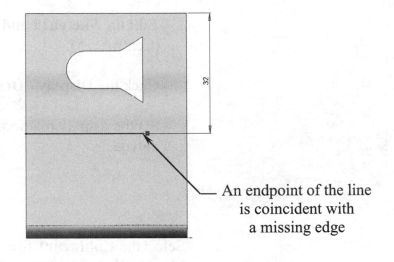

An endpoint of the line
is coincident with
a missing edge

- The endpoint of the line
changes to Blue color.
This indicates the sketch
is under defined.

- **Drag and drop** the
endpoint of the line until
it touches the vertical
right edge of the part.
A coincident relation
is added automatically.

- **Exit** the sketch or press
Control + Q.

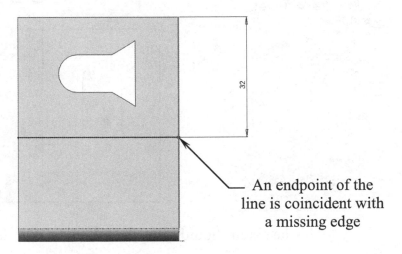

An endpoint of the
line is coincident with
a missing edge

- SOLIDWORKS continues to report the last error in the model. Click the **Close** the dialog box and continue to repair the error.

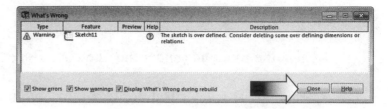

4. Repairing the 3rd error:

- The **Sketch11** has a plus sign next to its name; this indicates that the sketch is Over Defined.

- **Edit** the **Sketch11** and change to the Top orientation (Control + 5).

- Click the **Display / Delete Relations** button 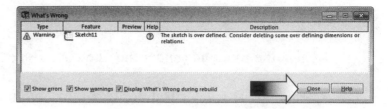.

- Set the display relations option to **Over Defining / Not-Solved**.

- Select the **Collinear1** from the list. This relation shows a **Magenta** color next to its Collinear symbol.

- **Delete** the **Collinear1** from the list.

- This last step should bring the sketch back to its Fully Define status.

- **Exit** the sketch or press **Control + Q**.

5. Saving your work:

- Click **File / Save As**.

- Enter **Repair Errors (Completed)** for the name of the file.

- Click **Save**.

- All errors have been repaired. The Feature-Manager tree is now free of errors.

- Close all documents.

Repair Errors & External References

1. Opening a part document:

- Browse to the Training Files folder.

- Open the document named: **Molded Part**.

- The **What's Wrong** dialog appears displaying the current errors in the model.

- Click **Close**. We are going to take a look at breaking the external references first.

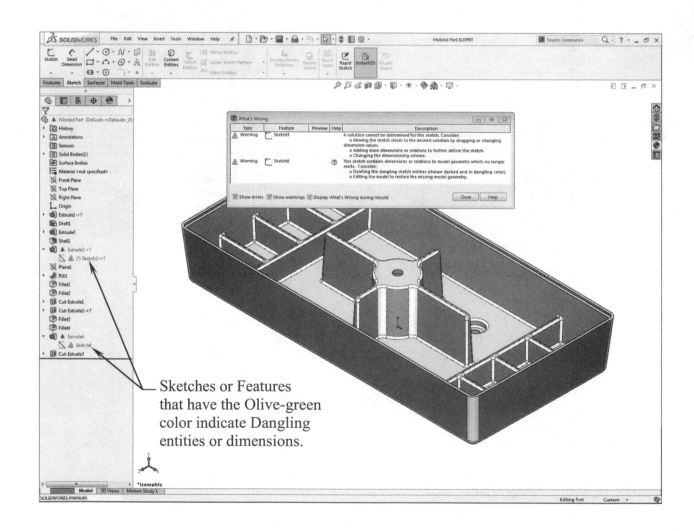

Sketches or Features that have the Olive-green color indicate Dangling entities or dimensions.

2. Breaking all External References:

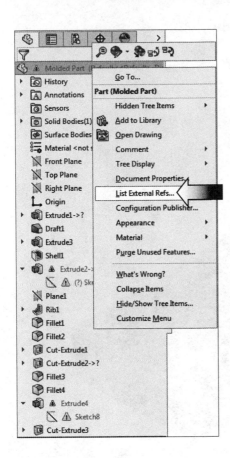

- Right click on the part's name and select: **List External Refs**.

- All of the Out of Context entities are displayed in the dialog box along with the names of the component parts in which they were related.

- Click **Break All** (Arrow).

- Click **OK** [OK] to confirm the deletion of all External-References.

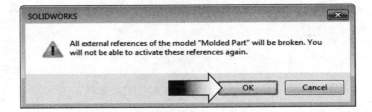

☀ *TIPS:*

4 basic steps should be done, in most cases, to repair or replace External References:

1. Break all External references (Right click on the part's name and select List External Refs.).

2. Replace the sketch Plane or Face (if missing).

3. Delete or replace any Relation with an External Reference symbol next to it (Display/Delete Relations).

4. Repair or replace the extrude type.

3. Replacing the Sketch Plane:

- Expand the **Extrude1** feature to see the **Sketch1** below.

- Right click on **Sketch1** and select **Edit Sketch Plane** .

- The Sketch Plane is missing; a new plane or face must be selected to replace it.

- Select the **Front** plane from the FeatureManager tree.

- After replacing the plane, click **OK**.

- The system displays the warning on other errors along with the solutions for repairing them.

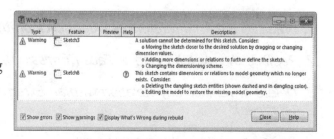

- Click **Close** .

4. Repairing the sketch Relations and Dimensions:

- Right click on **Sketch3** and select **Edit Sketch** .

- Click the **Display/Delete Relations** command or select:
Tools / Relations / Display-Delete.

- Delete the 2 geometric relations that have the external symbols next to their names (arrows).

- The sketch becomes fully-defined.

- Click **OK**.

- **Exit** the sketch .

5. Repairing the next sketch errors:

- Right click on **Sketch6** (under Cut-Extrude2) and select: **Edit Sketch** .

- Two dimensions are no longer attached to the part (circled).

Detached

Select the dimension to see the handle points. Drag the handle point of the dimension line (the red dot) to the model edge as shown, to re-attach it.

Re-attach the vertical dimension.

- The sketch becomes fully defined; the two dimensions are now attached to the edges of the model and locate the center of the circle.

- **Exit** the sketch .

- There is still a warning on the Cut-Extrude2.

6. Correcting the extrude type:

- Right click on **Cut-Extrude2** and select **Edit Feature**.

Select this face

- The surface that was used as the end condition option is no longer recognized; a new surface has to be selected for replacement.

- Select the **face** as indicated.

- Leave extrude depth as **.100**in. and the draft angle at **2deg**.

- Click **OK**.

7. Repairing the errors in the last sketch:

- Right click on the **Sketch8** and select **Edit Sketch**.

- Change to the Bottom orientation, or press **Control + 6** to switch to the bottom view.

Drag handle point to an edge to re-attach

- Dimensions created from or to other parts became Dangling and detached from the model (Olive-green color).

- Select one of the dimensions and drag its handle point (the red dot) to a model edge to re-attach.

- Repeat the same step to re-attach all other dimensions.

- The sketch becomes fully defined after all dimensions are re-attached.

- Change the dimension values as shown to reposition the text.

- **Exit** the sketch.

- The reference symbols and the error colors on the FeatureManager tree should now be all removed.

8. Saving your work:

- Select **File / Save As / Repair Errors / Save**.

Level 4: Final Exam

1. Open an existing part document:

<u>Go to:</u>

Training Files folder
Tooling Design folder
L4 Final Exam.sldprt.

2. Create a Parting Line:

- Direction of Pull = **TOP plane**.

- Draft Angle   = **1deg**.**

- Parting Lines   = **All outer edges** as shown.

Parting Lines —

** <u>*Important*</u> *- Roll back under the Loft feature and:*

a/ Add 1° drafts to all Yellow faces, including the 4 holes. This change will create some errors in the part.

b/ <u>Re-order</u> or <u>re-create</u> the fillets if necessary, after adding the drafts.

3. Create a Parting Surface:

- Use the **Perpendicular to Pull** option.

- For Parting Line: select the **Parting Line1** from the. FeatureManager tree.

4. Create an Offset Distance plane:

- Use the **Top** reference plane and **.375 in**. distance.

- The new plane is placed **above** the Top plane. (This new plane will also be used to create the Interlock Surfaces.)

5. Sketch the mold block:

- Sketch a **Rectangle** on the new plane (Plane1) and add dimensions* shown.

* Adjust the dimensions if needed to center the rectangle around the part.

- **Exit** the sketch.

6. Create a Tooling split:

- **1.250in**. (upper block)
- **1.250in**. (lower block)
- Use **Interlock Surface** with **5° Draft**.

7. Separate the mold blocks:

- Use the **Move/Copy** command to separate the two halves.

** <u>**OPTIONAL:**</u> Make the upper and lower solid bodies transparent for clarity.

8. Save your work as L4-Final.

Interlock Surfaces

Interlock Surfaces

CHAPTER 23

Using Appearances

A knurled feature would create hundreds of extra faces in the model and would increase file size, bogging down graphical performance. But if it is important to be able to represent a knurl then there are a couple of techniques that can be used to accommodate it.

This first half of the chapter will guide you through the method of modeling the Raised Diamond Knurls where a sketch profile will be swept along a path to create a spiral cut. The spiral cut is patterned and mirrored to repeat the raised diamond shapes of the knurl.

Modeling diamond knurls

1. Opening a part document:

- Browse to the Training Files folder and open a part document named: **Knurled Handle**.

- This model has a single solid body, a sweep profile, and a sweep path that was previously created.

Before   After

2. Creating a swept cut:

- Select the Features tool tab and click the **Swept Cut** command.

- For sweep profile, select the **Sketch2** from the FeatureManager tree.

- For Sweep path, select the **Helix**.

- Click **OK**.

- Rotate the model and examine the swept cut feature.

3. Creating a circular pattern:

- Click the **Circular Pattern** command under the Linear Pattern drop down arrow.

- Select the circular edge on the left end of the model for Pattern Direction.

- Enable the **Equal Spacing** checkbox (arrow).

- Enter **25** for Number of instances.

- Select the **Swept Cut** feature cither from the Feature tree or Directly from the graphics area.

- Click **OK**.

- Change to the top orientation (Control+5) to verify the pattern of the swept cut feature.

4. Creating a mirror pattern:

- Click the **Mirror** command from the Features tool tab.

- Expand the FeatureManager tree and select the **Front** plane to use as Mirror Plane.

- For Features to Mirror select both: the **Cut-Swept1** and the **CirPattern1** features.

- Enable the option:
  Geometry Pattern to
  create the pattern using only
  the geometry (faces and edges) of
  the features, rather than patterning and
  solving each instance of the feature.

- Click **OK**.

- Change to the top
  orientation (Control+5)
  to verify the result of the
  mirror pattern.

- **Save** and <u>close</u> the document.

Applying the knurl appearance

- Certain appearances require the use of RealView Graphics for more realistic Representation: knurled, dimpled, or sandblasted are some examples.

- RealView gives models a realistic and dynamic representation without the need to render. If your graphics card is RealView-compatible, RealView is enabled by default.

1. Opening a part document:

- Open a part document named:
Knurl Appearance.

- Click the drop-down arrow next to View Settings and enable the **RealView Graphics** option (arrow).

2. Applying the knurl appearance:

- Expand the **Task pane** on the right side of the screen and pin it.

- Expand the Metal and Steel folders. Drag and drop the **Stainless Steel Knurled** appearance onto the <u>surface</u> of the handle.

Drag & drop

- Select the **Apply to Face** option in the pop-up menu.

- The default knurled appearance is applied to the selected face.

- Next, we will modify the settings to make the diamond knurl appearance look a little more realistic.

3. Modifying the knurl appearance:

- <u>Right click</u> the surface of the handle where the knurled appearance was applied and select: **Appearance > Face** edit option (arrow).

- Select the **Advanced** and **Mapping** tabs (arrow).

- Select **Automatic** under the mapping drop down selection.

- Enter **.125in** for both **Width** and **Height** of the diamonds.

- Click **OK**.

- Rotate the model to verify the result of the knurls.

- Additionally, change the following:

* Mapping Type: **Cylindrical**

* Axis Direction: **ZX**

* Rotation: **90deg**.

- Click **OK**.

- **Save** and close the document.

Applying Wire Mesh appearance

- Similar to diamond knurls, modeling 3D wire mesh is just too time consuming, the rebuild time would be too long, and the amount of memory needed for the task is just simply not practical.

- Appearance once again can offer some acceptable results without sacrificing your computer performance, or the time it takes to create one.

- We will take a look at modifying an existing appearance and make it look like a wire meshed cable (pictured)

1. Opening a part document:

- Open a part document named: **Wire Mesh.sldprt**

- This model has 8 solid bodies and the sleeve in the middle will be used to apply the wire mesh.

2. Applying the wire mesh:

- Expand the **Task Pane** on the right side of the screen and pin it.

- Expand the following 2 folders: **Plastic > Composite** (arrows).

- Drag and drop the **Carbon Fiber Epoxy** appearance to the <u>middle sleeve</u> as noted.

Drag & drop

- Select the **Advanced** and the **Mapping** tabs and set the parameters shown below.

3. Modifying the wire mesh appearance:

- <u>Right click</u> the middle sleeve and select: **Appearance > Edit** (arrow).

- Select the **Advanced** and **Mapping** tabs. Set the Mapping type to Surface, and set the **Width** and **Height** to **.075in**, and the **angle** to **25deg**.

- Click **OK**.

- Alternatively, change the rotation to **90deg** (arrow) to change the angle of the mesh, if desired.

- **Save** and close the document.

Flatten Surfaces

This new feature is only available in SOLIDWORKS Premium.

You can flatten surfaces and multi-faced surfaces (such as surfaces that are split into multiple faces). Surfaces that have holes or other internal geometries cut out of the middle cannot be flattened.

1. Opening a part document:

    - Open the part document named:
     Flatten Surfaces.sldprt

    - Click **Insert > Surface > Flatten** .

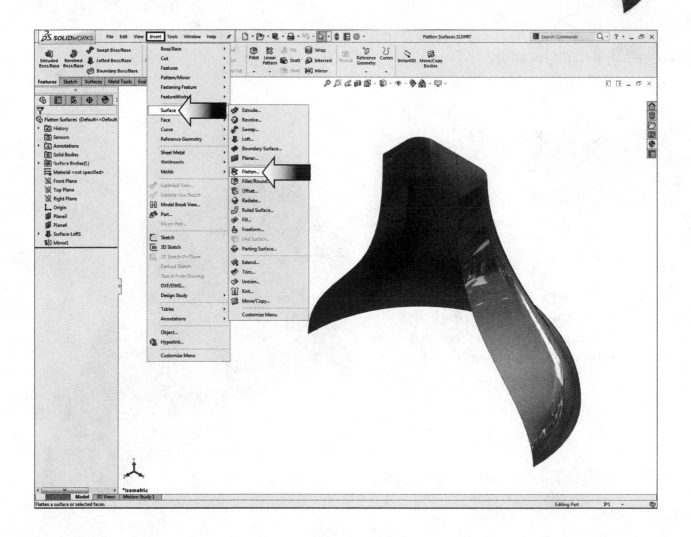

2. Flattening a surface:

- Click in the Face/ Surface to Flatten section and select the <u>left and right surfaces</u> of the model.

- Click in the Vertex to Flatten From and select <u>the vertex</u> as indicated.

Select vertex

Select 2 surfaces

- Leave the Accuracy slider at its default location.

- Click O**K**.

- The new flatten surface is created over the original surface. These surfaces can toggled to show or hide.

- Rotate the model to different angles to verify the result of the flatten surface.

3. Viewing the deformation plot:

- To view a deformation plot of the flattened surface, **right-click** the surface and click **Deformation Plot**.

- The deformation plot shows the areas on the flattened surface with the highest levels of stretch and compression. You can mouse over the surface to see the percent of deviation at any given point.

In the graphics area, right-click the flatten surfaces, and click Deformation Plot .

A deformation plot appears in the graphic area showing where the surface is compressed or stretched.

- Change to the front orientation and mouse over one of the red areas to see the percentage of the deformation (stretch).

4. Moving the surfaces:

- Click **Insert > Surface > Move/Copy**.

- In the Translate section, select the **Flatten Surface** to move.

- Enter **10.00in** in the Delta X box.

- Click **OK**.

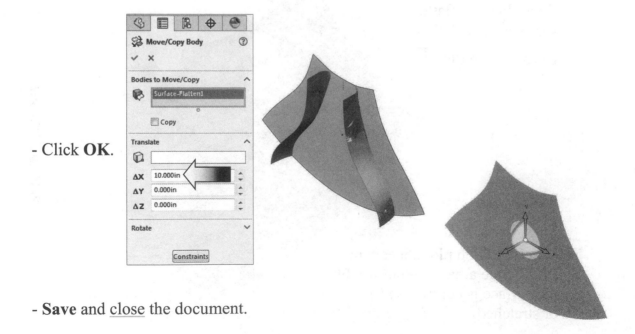

- **Save** and close the document.

Exercise: Flattening a Shoe Sole

1. Opening a part document:

- Open a Parasolid document named:
Shoe Sole.x_b

- Click **NO** to close the Import Diagnostic
dialog box.

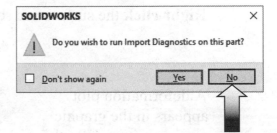

2. Flattening a surface:

- Click **Insert > Surface > Flatten** .

- For Surface to Flatten, select the **3 surfaces** of the entire sole.

- Press **F5** to display the Selection Filters at
the bottom left of the screen. Select the
Filter Vertices button (arrow).

- For **Vertex** to Flatten,
select the Vertex as noted.

Select vertex

- Enable the **Show Mesh Preview** and
Show Flat Preview checkboxes.

- Leave the Accuracy slider at its default value and click **OK**.

- Click off the Filter Vertices button and press F5 again to hide the Selection Filters toolbar.

- Click the **Surface-Imported1** on the FeatureManager tree and select **Hide** (arrow).

3. Viewing the deformation plot:

- **Right-click** the surface and select **Deformation Plot**.

- A deformation plot appears in the graphic area showing where the surface is compressed or stretched.

- Change to the front orientation (Control+1) and hover over one of the red areas to see the percentage of the deformation (stretch).

4. Saving your work:

- Click **File > Save As**.

- Enter **Shoe Sole Completed** for the file name.

- Press **Save**.

SOLIDWORKS 2019

Certified SOLIDWORKS Professional (CSWP)
Certification Practice for the Mechanical Design Exam

Courtesy of Paul Tran, Sr. Certified SOLIDWORKS Instructor

Certified SOLIDWORKS Professional (CSWP)
Certification Practice for the Mechanical Design Exam

Challenge I: Part Modeling & Modifications

Complete this challenge within 70 minutes (1 part)

(The following examples are intended to assist you in familiarizing yourself with the structures of the exams and the method in which the questions are asked.)

Create this part in SOLIDWORKS 2010 or newer
Unit: **Inches, 3 decimals**
Drafting Standards: **ANSI**
Origin: **Arbitrary**
Material: **Cast Alloy Steel**
Density: **0.264 lb/in^3**

(Save the model after each question
as a different file in case it must be
reviewed.)

(The CSWP-Segment 1 examines your skills on creating a parametric model, where various methods will be used to create and constraint the geometry of each features.)

1. Creating the solid bodies:

- Select the **Top** plane and open a new sketch.

- Sketch the profile shown below.

- Keep the Origin at the mid-point of the vertical line on the left side.

- Add the **construction lines**. Notice the two tapered construction lines end at .250" from the center of the rectangle.

- Add the **dimensions** to fully define the sketch (dimensions are in Inches, three decimal places).

2. Extruding the sketch:

- Switch to the **Features** tool tab and select the **Extruded Boss-Base** command.

- For Direction 1, select the **Blind** type.

- For Extrude Depth, enter **1.500in**

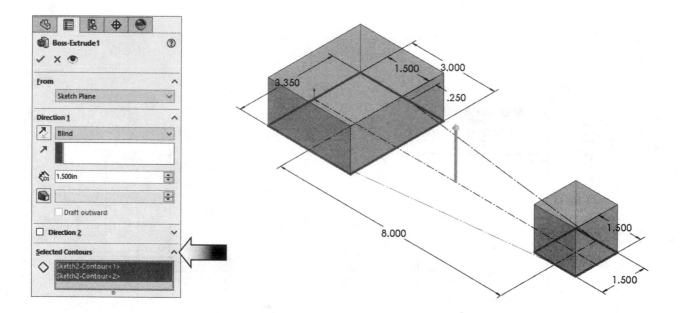

- Expand the **Selected Contour** section (arrow) and select the **2 rectangles** in the graphics area (by clicking 1 line in each rectangle the whole profile will be selected).

- The preview graphics should show 2 rectangles are being extruded at the same time.

- Click **OK**.

3. Trimming the solid bodies:

- Open a new sketch on the **Front** plane.

- Sketch **2 rectangles** and make the corners Coincident to the 2 solid bodies.

- Add the **dimensions** shown to fully define the sketch.

4. Extruding a cut:

- Switch to the **Features** tool tab and select the **Extruded Cut** command.

- For Direction 1, select **Through-All Both**.

- Expand the **Feature Scope** section (arrow).

- Select the option:
  All Bodies (arrow).

- Click **OK**.

- A multibody part is created. The 2 solid bodies are not connected at this point.

5. Adding the bridge feature:

- Select the **Front** plane and open a new sketch.

- Sketch the profile shown here.

- <u>Show</u> the **Sketch1** under the Boss-Extrude1 (arrow).

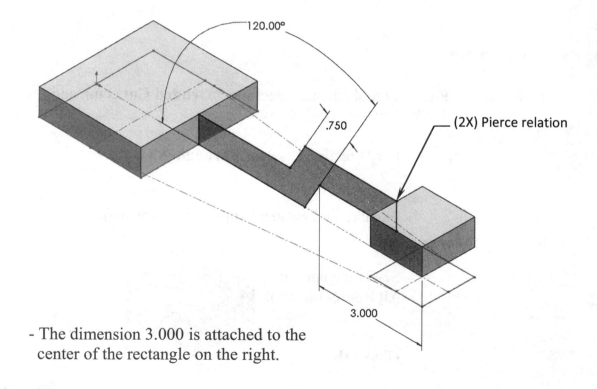

- The dimension 3.000 is attached to the center of the rectangle on the right.

- Add the relations as indicated to fully define the sketch.

6. Extruding the bridge:

- Switch to the **Features** tool tab and select the **Extruded Boss-Base** command.

- For Direction 1, select the **Mid Plane** type.

- For Extrude Depth, enter **.375in**

- Enable the **Merge Result** checkbox (arrow)

- Expand the **Feature Scope** section and select the **All Bodies** option.

- Click **OK**.

- The bridge feature joins all 3 solid bodies into one single solid body at this point.

7. Assigning names to the main dimensions:

- Each dimension has a default name such as D1, D2, D3, etc.
The main dimensions will get changed to Alphabetical order like A, B, C, etc.

- Right click the **Annotation** folder (arrow) and enable the **Display Annotations** option (arrow).

- Select **View**, **Hide/Show**, **Dimension Names** (arrow).

- Both dimensions and dimension names are displayed in the graphics area.

- Select the dimension **8.000** (circled).

- Locate the **Primary Value** section on the FeatureManager tree.

- Change the **D1** value to **A** (arrow). It is important to keep the **@Sketch1** the
  same; this is a link to the actual sketch. Do not modify it.

- Select the dimension **3.000** (circled).

- Change the default name **D1** (or D2) to **B** (arrow).

- The new names A & B should appear under the two dimensions.

- Change to the **Top** view orientation (Control + 5).

- Select the dimension **1.500** (circled).

- Change the default name to **C** (arrow).

- Change to the **Isometric** view orientation (Control + 7).

- Select the dimension **1.500** (circled).

- Change the default name to **D** (arrow).

- There should be 4 letters assigned to 4 dimensions at this point:
  A=8.000, B=3.000, C=1.500, and D=1.5000

- The last dimension is the **3.000** (circled). Change its name to **E** (arrow).

8. Creating equations and links:

- Equations and links can help to speed up the process of making changes to multiple features in a model.

- Select **Tools, Equations**. Press **Control+5** (Top view).

- The 1st dimension will be the **Driven** dimension and the 2nd dimension will be the **Driving** dimension.

- Click in the cell under Equations (arrow).

- Select the 1st dimension **1.500** (**C**) marked with number 1 in a circle.

- The equal sign is added to the Value / Equation cell.

- Select the 2nd dimension **3.000 (B)**. It is marked with a number 2 in a circle. This will be the Driving dimension.

- Enter **/2** (divided by 2).

- The equation should look something like this: **"C@Sketch1" = "B@Sketch1" / 2**

- The Evaluates to and the Comments cells are filled in automatically.

<u>Repeat:</u>

- Click in the next cell below Equation (arrow).
Press **Control+7** (Isometric view).

- Select the dimension **1.500 (D)**; it is marked with a number 1 in a circle. This will be the Driven dimension.

- The Equal sign is added.

- Select the dimension **3.000 (B)**; it is marked with a number 2 in a circle. This dimension will be the Driving dimension.

- Enter **/2** (divided by 2).

- The Evaluates-to and the Comments cells are filled in automatically.

- The equation should look like this: **"D@Boss-Extrude1" = "B@Sketch1" / 2**

9. Finding the mass of the part:

- For clarity, let's hide the dimensions and annotations. Right click the Annotations folder and click-off the Display Annotation option.

- Ensure that the material is set to: **Cast Alloy Steel.**

- Switch to the **Evaluate** tab.

- Click **Mass Properties**.

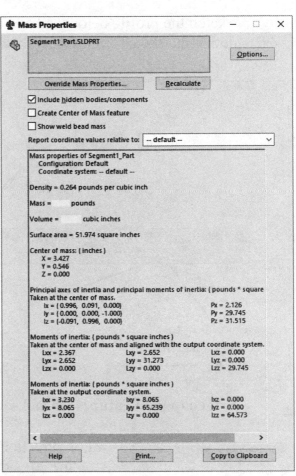

- Enter the mass of the part here:

_____ pounds.

10. Saving the part:

- Save the model after each question as a different file in case it must be reviewed.

- Select **File, Save As**.

- Enter **Segment1_Q1** for the file name.

- Click **Save**.

11. Adding the support base:

- The next step is to add the support base across the width or the part.

- Select the **Front** plane and open a new sketch.

- Sketch the profile below and add the dimensions/relations indicated to fully define the sketch. (Show the Sketch1 to add the Coincident relations.)

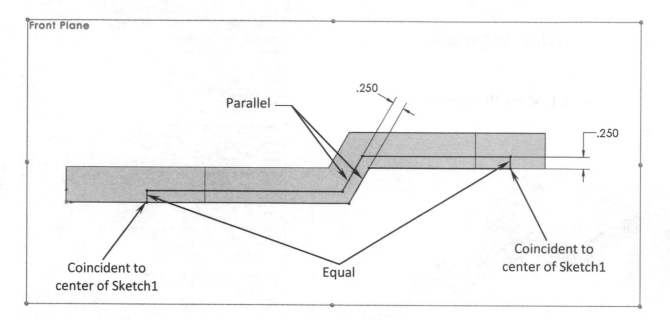

- Switch to the **Features** tab and select the **Extruded Boss-Base** command.

- For Direction 1, select the **Up to Surface** and click the **face** on the left side of the model as noted.

- For Direction 2, select the **opposite face** as indicated.

- Click **OK**.

12. Trimming the support base:

- Open a new sketch on the **Top** plane.

- Sketch the profile shown. Use the Mirror command to keep the 2 halves symmetrical with each other.

- Switch to the **Features** tab and select **Extruded Cut**.

- For Direction 1, select **Through All**.

- Click the **Reverse** button (arrow).

- Click **OK**.

13. Calculating the mass properties:

- Switch to the **Evaluate** tab.

- Click **Mass Properties**.

- Enter the mass of the part here:

_____ pounds.

14. Saving the part:

- Select **File, Save As**.

- Enter **Segment1_Q2** for the file name.

- Click **Save**.

- Keep the part open.

15. Modifying the dimensions:

- Right click the **Annotations** folder and enable the **Display Annotations** option.

- Change the dimension **8.000** (circled) to **9.000**.

- Change the dimension **3.000** (circled) to **3.500**.

- Press **Rebuild** ⬛ to update the model.

16. Calculating the mass:

- Right click the **Annotations** folder and click-off the Display Annotations option.

- Switch to the **Evaluate** tab.

- Click **Mass Properties.**

- Enter the mass of the part here:

_____ pounds.

17. Saving the part:

- Select **File, Save As.**

- Enter **Segment1_Q3** for the file name.

- Click **Save.**

- Keep the part open.

18. Adding 6-sided polygon holes:

- Select the **Top** plane and open a new sketch.

- Sketch **2 Polygons** and add the diameters dimensions to the construction circles. (Show the Sketch1. The centers of the polygons are Coincident to the centers of the rectangles in Sketch1.)

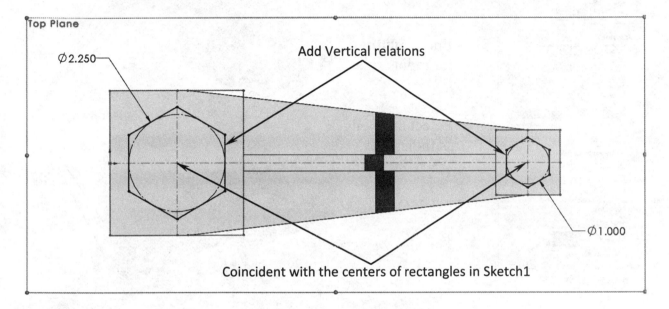

- Add a **Vertical** relation to one of the lines in each polygon.

- Switch to the **Features** tab and click **Extruded Cut**.

- For Direction 1, select **Through All**.

- Click the **Reverse** button (arrow).

- Click **OK**.

19. Adding fillets:

- Click the **Fillet** command on the **Features** tab.

- Use the default **Constant Size Fillet** option.

- Enter **.500in** for radius size.

- Select the **8 edges** as indicated.

Select 8 edges

- Click **OK**.

- Select the **Fillet** command once again.

- Use the default **Constant Size Fillet** option.

- Enter **.250in** for radius size.

Select 12 edges

- Select the **12 edges** as noted.

- Click **OK**.

20. Calculating the mass:

- Switch to the **Evaluate** tab.

- Click **Mass Properties.**

- Enter the mass of the part here:

_____ pounds.

21. Saving the part:

- Select **File, Save As**.

- Enter **Segment1_Q4** for the file name.

- Click **Save**.

- Keep the part open.

22. Modifying the dimensions:

- Right click the **Annotations** folder and enable the **Display Annotations** option.

- Change the dimension **3.500** (circled) to **3.125in**.

- Change the dimension **9.000** (circled) to **8.250in**.

- Press **Rebuild** to update the model.

23. Calculating the mass:

- Right click the **Annotations** folder and click-off the Display Annotations option.

- Switch to the **Evaluate** tab.

- Click **Mass Properties.**

- Enter the mass of the part here:

_____ pounds.

24. Saving the part:

- Select **File, Save As**.

- Enter **Segment1_Q5** for the file name.

- Click **Save**.

- Keep the part open.

25. Creating the side tabs:

- Select the **Front** plane and open a new sketch.

- Sketch the profile and add the dimensions / relations to fully define the sketch. (Use the Sketch Fillet to create the R.250 arc.)

- Switch to the **Features** tab and select the **Extruded Boss-Base** command.

- For Direction 1, select the **Up-to-Surface** option and click the face on the left side of the part as indicated.

- For Direction 2, also select the **Up-to-Surface** option and click the face on the right side of the part.

- Click **OK**.

26. Adding a through hole:

- Open a new sketch on the **Front** plane.

- Sketch a **Circle** and make it **Concentric** with the R.250 radius.

- Add the **.188 diameter** dimension to fully define the sketch.

- Click **Extruded Cut** on the Features tab.

- For Direction 1, select **Through All – Both**.

- Click **OK**.

27. Calculating the mass:

- Switch to the **Evaluate** tab.

- Click **Mass Properties**.

- Enter the mass of the part here:

_____ pounds.

28. Saving the part:

- Select **File**, **Save As**.

- Enter **Segment1_Q6** for the file name.

- Click **Save**.

- Close the part document.

Certified SOLIDWORKS Professional (CSWP)
Certification Practice for the Mechanical Design Exam

Challenge II: Part Modifications & Configurations

Complete this challenge within 50 minutes (3 parts)

(The following examples are intended to assist you in familiarizing yourself with the structures of the exams and the method in which the questions are asked.)

Modify this part using SOLIDWORKS 2010 or newer
Unit: **Inches, 3 decimals**
Drafting Standards: **ANSI**
Origin: **Arbitrary**
Material: **ABS**
Density: **0.2037 lb/in^3**

(This portion of the test will examine your skills on the modification of dimensions, geometry, and repair errors in a part.)

1. Opening a part document: (1 of 3)

- Click **File / Open**.

- Locate and open the part document named: **Segment 2A.sldprt**

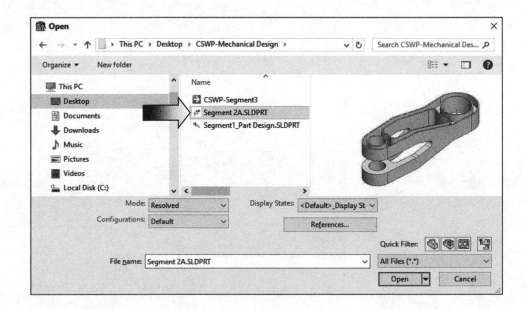

2. Modifying dimensions:

- Double click the feature named **Cut-Extrude1** and change the dimension **4.875** (circled) to **4.000**.

- Click **Rebuild** to update the model.

- Before the dimension change.

- After the dimension change.

- Double click the **Shell** feature from the FeatureManager tree.

- Change the wall thickness dimension from **.090** to **.125**.

- Click **Rebuild** to update the model.

3. Calculating the mass:

- Switch to the **Evaluate** tab.

- Click **Mass Properties**.

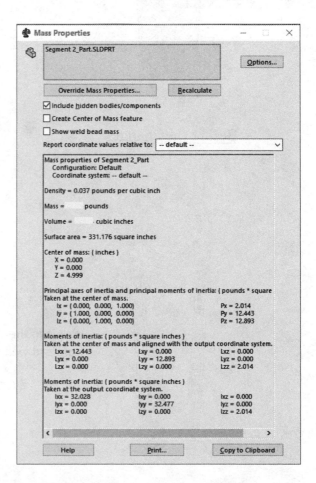

- Enter the mass of the part here:

_____ pounds.

4. Modifying the geometry:

- <u>Edit</u> the **Sketch1** under the Boss-Extrude1 feature.

- Modify the geometry of the sketch so that it would look like the images on the right.

<u>Before the changes</u>

<u>After the changes</u>

Remove the hole and wall

Extend wall and make Concentric

5. Calculating the mass:

- Switch to the **Evaluate** tab.

- Click **Mass Properties**.

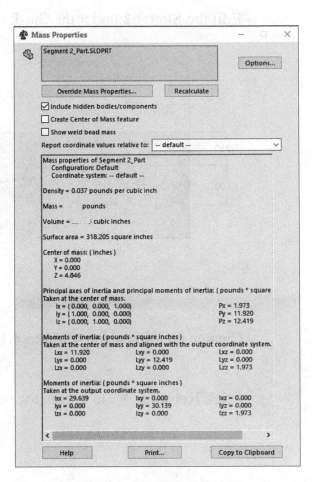

Mass Properties

Segment 2_Part.SLDPRT

Options...

Override Mass Properties...   Recalculate

☑ Include hidden bodies/components
☐ Create Center of Mass feature
☐ Show weld bead mass

Report coordinate values relative to: -- default --

Mass properties of Segment 2_Part
   Configuration: Default
   Coordinate system: -- default --

Density = 0.037 pounds per cubic inch

Mass =   pounds

Volume = _____ cubic inches

Surface area = 318.205 square inches

Center of mass: (inches)
   X = 0.000
   Y = 0.000
   Z = 4.846

Principal axes of inertia and principal moments of inertia: (pounds * square
Taken at the center of mass.
   Ix = (0.000, 0.000, 1.000)   Px = 1.973
   Iy = (1.000, 0.000, 0.000)   Py = 11.920
   Iz = (0.000, 1.000, 0.000)   Pz = 12.419

Moments of inertia: (pounds * square inches)
Taken at the center of mass and aligned with the output coordinate system.
   Lxx = 11.920   Lxy = 0.000   Lxz = 0.000
   Lyx = 0.000   Lyy = 12.419   Lyz = 0.000
   Lzx = 0.000   Lzy = 0.000   Lzz = 1.973

Moments of inertia: (pounds * square inches)
Taken at the output coordinate system.
   Ixx = 29.639   Ixy = 0.000   Ixz = 0.000
   Iyx = 0.000   Iyy = 30.139   Iyz = 0.000
   Izx = 0.000   Izy = 0.000   Izz = 1.973

Help   Print...   Copy to Clipboard

- Enter the mass of the part here:

_____ pounds.

6. Making additional changes:

- <u>Edit</u> the **Sketch1** under the Boss-Extrude1 feature.

- Change the dimensions to match the values in the circles.

- Add any relations needed to fully define the sketch.

- **Exit** the sketch and **Rebuild** the model. The model should not have any errors. Correct the errors when/where they occur.

R2.500
∅3.800
R.375
R3.000
12.000
.600
∅1.550
R1.375

- <u>Edit</u> the **Sketch2** under the Cut-Extrude1 feature.

- Change the dimension **4.250** to **4.750** (circled).

- Click **Rebuild** to update the model.

- Double click the **Shell** feature from the FeatureManager tree.

- Change the wall thickness dimension from **.125** to **.100**.

- Click **Rebuild** to update the model.
  The model should be free of errors. Correct/repair any errors when/where they occur.

7. Calculating the mass:

- Switch to the **Evaluate** tab.

- Click **Mass Properties**.

- Enter the mass of the part here:

_____ pounds.

8. Saving the part:

- Select **File, Save As**.

- Enter **Segment 2A (Completed).sldprt** for the file name.

- Click **Save**.

- Close the part document.

1. Opening a part document: (2 of 3)

- Click **File / Open**.

- Locate and open the part document named: **Segment 2B.sldprt**

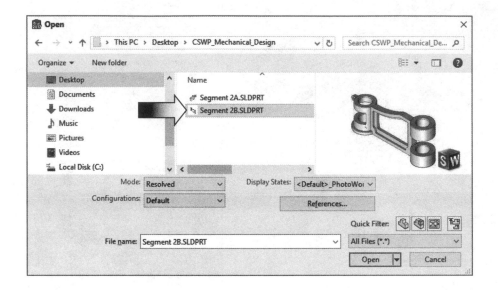

(This portion of the test will examine your skills on the modification of dimensions, geometry, and repair errors in a part.)

2. Modifying dimensions:

- Double click the feature **Revolve2** from the Feature-Manager tree.

- Change the dimension **8.750** (circled) to **8.500**.

- Click **Rebuild** to update the model.

- The dimension change triggers some errors in the model. These errors will be repaired after the geometry in the Sketch3 is corrected.

- <u>Edit</u> the **Sketch3** under the Boss-Extrude1 feature (arrow).

- The dimension 12.250 prevents the line from being tangent to the arc. Delete the dimension **12.250** as indicated.

- The sketch is no longer over defined.
  Add a **Horizontal** relation to the line as noted.

- <u>Exit</u> the sketch when it is fully defined.

- The errors are automatically corrected and removed.

3. Suppressing features:

- Locate the feature named **Cut-Extrude3** from the FeatureManager tree. It is the cutout feature in the middle of the part.

- Click the Cut-Extrude3 feature and select **Suppress**.

- The cutout feature is suppressed.

4. Calculating the mass:

- Switch to the **Evaluate** tab.

- Click **Mass Properties**.

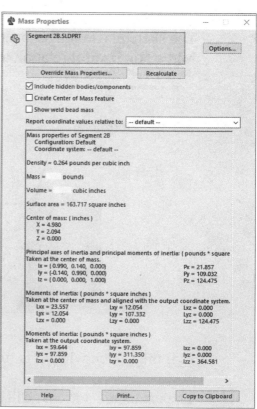

- Enter the mass of the part here:

_____ pounds.

5. Changing a dimension linked to an equation:

- Right click the Annotations folder and select **Display Annotations**.

- The dimension 3.500 (circled) is tied to the dimension 5.000 (circled) in an equation that was previously created. It is the driving dimension.

- Double click the dimension **3.500** and change it to **4.000**.

- Click **Rebuild** to update the model. Hide all annotations.

6. Modifying the revolved feature:

- Select the **Top** plane and open a new sketch.

- Select the <u>outer edge</u> of the revolved feature (on the right side) and create an **Offset Entities** using a distance of **.300in**. (smaller). Also **Convert** the <u>outer edge</u> of the same revolved feature into a circle as indicated.

Convert entity

.300

Offset entity

- Switch to the **Features** tab and click **Extruded Cut**.

- For **Extrude From**, select **Offset** and enter **.400**

- For **Direction 1**, select **Offset From Surface** (arrow)

Select face

- Select the <u>face</u> noted to offset from.

- Enter **.400in** for distance.

.300

- Click **OK**.

- The cut feature causes the part to have 2 disjoined solid bodies. The gap between the 2 bodies must be filled in.

Disjoined

7. Filling the gap:

- Select the **face** as indicated and open a new sketch.

- Select the <u>outer edge</u> of the cylinder and click **Convert Entities** as noted.

Convert entity

Sketch face

- Add 3 additional **lines** as shown below.

- Trim the circle and add the **1.500** dimension.

- Add **Collinear** relations between the lines and the model edges.

Add 3 lines

(2X) Collinear to edge

1.500

- Switch to the **Features** tab and select **Extruded Boss-Base**.

- For Direction 1, select **Up-to-Surface** and click the <u>face</u> as noted.

- For Direction 2, also select **Up-to-Surface** and click the <u>face</u> indicated.

- Click **OK**.

- Rotate the model and examine the extruded feature from different orientations.

8. Calculating the mass:

- Switch to the **Evaluate** tab.

- Click **Mass Properties**.

- Enter the mass of the part here:

_____ pounds.

9. Making the final changes:

- Change the dimension **8.500** in the <u>Revolve2</u> to **9.000** (circled).

- Change the dimension **4.000** in the <u>Revolved1</u> to **3.400** (circled).

- Change the dimension **.400** in the <u>Revolved1</u> to **.300** (circled).

- Also change both of the dimension **.400** in the <u>Cut Extrude4</u> to **.300** (circled).

10. Calculating the mass:

- Switch to the **Evaluate** tab.

- Click **Mass Properties**.

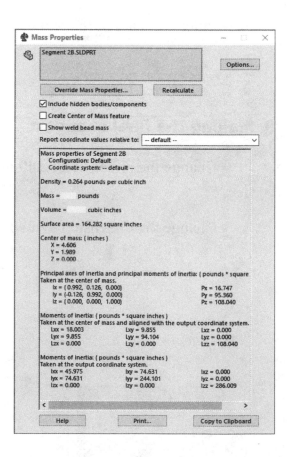

- Enter the mass of the part here:

_____ pounds.

- Save your model as **Segment 2B (Completed)** and close the part document.

1. Opening a part document: (3 of 3)

- Click **File / Open**.

- Locate and open the part document named: **Segment 2C.sldprt**

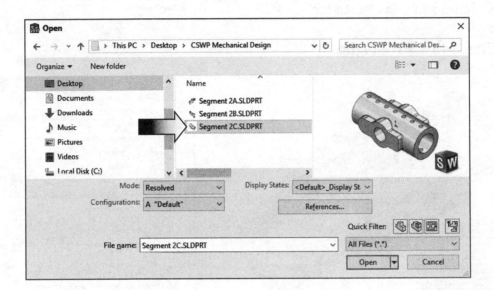

(This portion of the test will examine your skills on the modification of existing configurations, as well as creating new configurations in a part.)

2. Locating the configurations:

- Click the **Configurations** tab (arrow).

- How many configurations are there in this model?

- Enter the number of configurations found here:

_____ (1, 2, or 3).

3. Calculating the mass:

- Double click configuration **B** to activate it. Each configuration has its own material and different feature sizes.

- Switch to the **Evaluate** tab.

- Click **Mass Properties**.

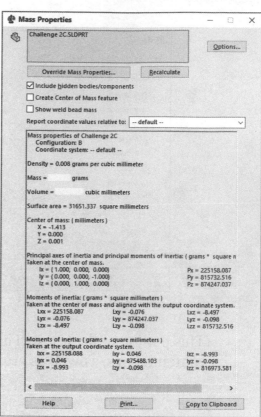

- Enter the mass of configuration **B** here:

_____ pounds.

4. Adding a hole:

- Double click configuration **C** to activate it.

- Open a new sketch on the <u>face</u> as noted.

- Sketch a **circle** and add a diameter dimension to fully define the sketch.

Sketch face

Ø9.50

Coincident

- Switch to the **Features** tab and click **Extruded Cut**.

- For Direction 1, select **Through All**.

- Click **OK**.

5. Calculating the mass:

- Switch to the **Evaluate** tab.

- Click **Mass Properties**.

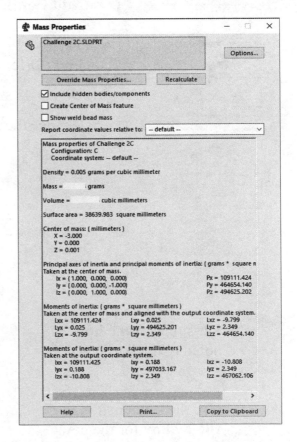

- Enter the mass of configuration **C** here:

_____ pounds.

6. Creating a new configuration:

- Double click configuration **A** to activate it.

- Right click anywhere inside the configuration tree and select:
  Add Configuration.

- Enter **D** for the name of the new configuration.

- Change the **3 dimensions** to match the values in the circles.

- Un-suppress the **hole** as noted (Cut-Extrude4).

- Use the same material **1060 Alloy** for the new configuration.

- The hole (Cut-Extrude4) should be <u>unsuppressed</u> in the configurations C and D.

- The hole (Cut-Extrude4) should be <u>suppressed</u> in the configurations A and B.

7. Calculating the mass:

- Remain in configuration **D**.

- Switch to the **Evaluate** tab.

- Click **Mass Properties**.

- Enter the mass of configuration **D** here:

_____ pounds.

8. Saving your work:

- Select **File, Save As**.

- Enter: **Segment 2C (Completed)** for the file name.

- Click **Save**.

- Close all part documents.

Certified SOLIDWORKS Professional (CSWP)
Certification Practice for the Mechanical Design Exam

Challenge III: Bottom Up Assembly & Mates

Complete this challenge within 80 minutes

(The following examples are intended to assist you in familiarizing yourself with the structures of the exams and the method in which the questions are asked.)

Modify this part using SOLIDWORKS 2010 or newer
Unit: **Inches, 3 decimals**
Drafting Standards: **ANSI**
Origin: **Arbitrary**
Material: **Already Specified for each part**

(This portion of the test will examine your skills on constraining components using the Bottom Up Assembly approach.)

1. Starting a new assembly document:

- Click **File / New**.

- Select an **Assembly Template** and click **OK**.

- Change the Drafting Standard to **ANSI** and the Units to **IPS**, 3 decimal places.

- You will be asked to demonstrate the use of different mate types to constrain the component in an assembly.

2. Inserting the 1st component:

- Change to the **Assembly** tab and select: **Insert Component**.

- Locate the component named **Base.sldprt** and click **Open**.

- Click the **Green check** to place the component on the Origin.

3. Inserting other components:

- Insert all other components into the assembly.

- Place the components approximately as shown. Make a copy of the Link1 and Link2.

- Rotating the Claw and the Link ahead of time will make it easier to see and to mate them to the Base.

- Also hide the origins.
  Click **View**, **Hide/Show**, **Origins**.

4. Mating the components:

- Use the **Filter Faces** (press the letter **X** shortcut key) to help select the faces more easily.

- Hold the **Control** key and select the <u>hole</u> and the <u>shaft</u> as noted.

- Release the control key to see the mate shortcuts pop-up options.

Concentric

- Select **Concentric** (arrow) and click the **Green** check (OK).

- Add a **Coincident** mate between the **back face** of the <u>Pivot Plate</u> and the **front face** of the <u>Base</u>.

Coincident

- Also add a **Parallel** mate between the **side face** of the <u>Pivot Plate</u> and the **side face** of the <u>Base</u>.

Parallel

- Add a **Concentric** mate between the **bottom hole** of the <u>Link1</u> and the **bottom pin** on the left side of the <u>Base</u>.

- Also add a **Concentric** mate between the **bottom hole** of the <u>Link2</u> and the **top pin** on the left side of the <u>Base</u>.

(2X) Concentric
Pin & Hole

(2X) Coincident
Front & back

- Add a **Coincident** mate between the **back face** of the <u>Link1</u> and <u>Link2</u> and the **front face** of the <u>Base</u>.

5. Assembling the 1ˢᵗ Claw:

- Zoom closer to the Claw and the Links components.

- Create a **Concentric** mate between the **lower hole** of the <u>Claw</u> and the **upper hole** of the <u>Link1</u>.

Concentric

Concentric

- Create another **Concentric** mate between the **upper hole** of the <u>Claw</u> and the upper hole of the **Link2**.

- Add a **Width** mate between the **front** and the **back faces** of the <u>Claw</u> and one of the <u>Links</u> (total of 4 faces).

Face 3 & 4

Width

Face 1 & 2

- Change to the **Right** orientation (Control+4).

- Zoom very close and examine the gaps between the Claw and the 2 Links.
  The gaps must be exactly the same on both sides.

Right Side view

Gaps

Coincident (no gap)

Coincident (no gap)

- Double check the back side of the 2 Links should be Coincident to the Pivot
  Plate.

- The back side of the Pivot Plate should also be Coincident to the Base.

6. Assembling the 2ⁿᵈ Claw:

- Add a **Concentric** relation and a **Coincident** relation between the <u>Link1</u> and the **Lower Pin** in the <u>Base</u>.

- Add the same **Concentric** and **Coincident** relations between the <u>Link2</u> and the **Upper Pin** in the <u>Base</u>.

(2X) Concentric & Coincident
Pin & Hole

- Also add a **Width** mate between the **Claw2** and one of the **Links**. (Use the front & back faces of the Claw2, and the front & back faces of the Link.)

- Move the 2 claws back and forth to ensure that they can be moved freely.

7. Detecting Collision:

- Switch to the **Assembly** tab and click the **Move Component** command.

- Expand the **Options** section and select the **Collision Detection** option (arrow).

- Enable the **Stop at Collision** checkbox.

- Enable the **Highlight Faces** check-box

- Drag the Claw2 towards the center. It should stop when the 2 Links collided.

The 2 Links collided here

- Click **OK** but do not move the Claw2.

8. Measuring the angle:

- Switch to the **Evaluate** tab.

- Select the **narrow-face** on the side of the Claw2 and the **Right** plane of the Assembly.

- Locate the **Angle Measurement** and enter it here:

_____ degrees.

Select Face of Claw2 and the Right plane

Face<1> @Claw1-2
Right Plane

Angle:
Distance: 1.155in
Delta X: 1.155in
Delta Y: 0.000in
Delta Z: 0.000in

File: Claw1-2@Segment 3 Assembly To: Segment 3
Assembly.SLDASM
File: Segment 3 Assembly.SLDASM Config: Default

9. Adding a Symmetric mate:

- Add a **Symmetric** mate using the **Right** plane of the <u>Assembly</u> and the **2 small faces** of the <u>2 Claws</u>.

Select Right plane & 2 faces

- Click **OK**.

- Drag one of the Claws back and forth to ensure that it can move freely.

10. Moving components with Collision Detection:

- Click **Move Component** .

- Select the following options:

  * Collision Detection

  * These Components (select the 2 Claws)

  * Stop at Collision

  * Highlight Faces

- Click the **Resume Drag** button and drag one of the claws until it stops.

- It is important to keep the Claws at the collided position for measurement.

Drag until stops

11. Calculating the Center of Mass:

- Switch to the **Evaluate** tab.

- Click **Mass Properties**.

- Enter the **Center of Mass** here:

X = _____

Y = _____

Z = _____

12. Adding an angle mate:

- Rotate the assembly to see the right side of the Link2 component.

- Add a **45°** mate between the **face** on the right side of the <u>Link1</u> and the **Right** plane of the <u>Assembly</u>.

- This mate should fully define the assembly and all components.

13. Calculating the Center of Mass:

- Switch to the **Evaluate** tab.

- Click **Mass Properties**.

- Enter the **Center of Mass** here:

X = _____

Y = _____

Z = _____

14. Modifying the Mass:

- This step will modify the values of the Mass, Center of mass, and Moments of Inertia of the model.

- Click **Mass Properties**.

- Click the **Override Mass Properties** button (arrow).

- Enable the **Override Mass** checkbox.

- Enter **15.258lbs**.

- Enable the **Override Center of Mass** checkbox.

- Enter the following:

    X = -44.45in
    Y = -44.50in
    Z = -101.60in

- Click **OK**.

15. Calculating the modified Center of Mass:

- Switch to the **Evaluate** tab.

- Click **Mass Properties**.

- Enter the modified **Center of Mass** here:

    X = _____

    Y = _____

    Z = _____

16. Saving your work:

- Select **File**, **Save As**.

- Enter: **Segment 3C (Completed)** for the file name.

- Click **Save**. Close all part documents.

Glossary

Alloys:

An Alloy is a mixture of two or more metals (and sometimes a non-metal). The mixture is made by heating and melting the substances together.
Examples of alloys are Bronze (Copper and Tin), Brass (Copper and Zinc), and Steel (Iron and Carbon).

Gravity and Mass:

Gravity is the force that pulls everything on earth toward the ground and makes things feel heavy. Gravity makes all falling bodies accelerate at a constant 32ft. per second (9.8 m/s). In the earth's atmosphere, air resistance slows acceleration. Only on airless Moon would a feather and a metal block fall to the ground together.
The mass of an object is the amount of material it contains.
A body with greater mass has more inertia; it needs a greater force to accelerate.
Weight depends on the force of gravity, but mass does not.

When an object spins around another (for example: a satellite orbiting the earth) it is pushed outward. Two forces are at work here: Centrifugal (pushing outward) and Centripetal (pulling inward). If you whirl a ball around you on a string, you pull it inward (Centripetal force). The ball seems to pull outward (Centrifugal force) and if released will fly off in a straight line.

Heat:

Heat is a form of energy and can move from one substance to another in one of three ways: by Convection, by Radiation, and by Conduction.

- Convection takes place only in liquids like water (for example: water in a kettle) and gases (for example: air warmed by a heat source such as a fire or radiator). When liquid or gas is heated, it expands and becomes less dense. Warm air above the radiator rises and cool air moves in to take its place, creating a convection current.

- Radiation is movement of heat through the air. Heat forms match set molecules of air moving and rays of heat spread out around the heat source.

- Conduction occurs in solids such as metals. The handle of a metal spoon left in boiling liquid warms up as molecules at the heated end move faster and collide with their neighbors, setting them moving. The heat travels through the metal, which is a good conductor of heat.

Inertia:

A body with a large mass is harder to start and also to stop. A heavy truck traveling at 50mph needs more power breaks to stop its motion than a smaller car traveling at the same speed.

Inertia is the tendency of an object either to stay still or to move steadily in a straight line, unless another force (such as a brick wall stopping the vehicle) makes it behave differently.

Joules:

The Joule is the SI unit of work or energy.

One Joule of work is done when a force of one Newton moves through a distance of one meter. The Joule is named after the English scientist James Joule (1818-1889).

Materials:

- Stainless steel is an alloy of steel with chromium or nickel.

- Steel is made by the basic oxygen process. The raw material is about three parts melted iron and one part scrap steel. Blowing oxygen into the melted iron raises the temperature and gets rid of impurities.

- All plastics are chemical compounds called polymers.

- Glass is made by mixing and heating sand, limestone, and soda ash. When these ingredients melt they turn into glass, which is hardened when it cools. Glass is in fact not a solid but a "supercooled" liquid. It can be shaped by blowing, pressing, drawing, casting into molds, rolling, and floating across molten tin, to make large sheets.

- Ceramic objects, such as pottery and porcelain, electrical insulators, bricks, and roof tiles are all made from clay. The clay is shaped or molded when wet and soft, and heated in a kiln until it hardens.

Machine Tools:

Are powered tools used for shaping metal or other materials, by drilling holes, chiseling, grinding, pressing or cutting. Often the material (the work piece) is moved while the tool stays still (lathe), or vice versa, the work piece stays still while the tool moves (mill). Most common machine tools are Mill, Lathe, Saw, Broach, Punch press, Grind, Bore and Stamp break.

CNC

Computer Numerical Control is the automation of machine tools that are operated by precisely programmed commands encoded on a storage medium, as opposed to controlled manually via hand wheels or levers, or mechanically automated via cams alone. Most CNC today is computer numerical control in which computers play an integral part of the control.

3D Printing

All methods work by working in layers, adding material, etc. different to other techniques, which are subtractive. Support is needed because almost all methods could support multi material printing, but it is currently only available in certain top tier machines.

A method of turning digital shapes into physical objects. Due to its nature, it allows us to accurately control the shape of the product. The drawback is size restraints and materials are often not durable.

While FDM doesn't seem like the best method for instrument manufacturing, it is one of the cheapest and most universally available methods.

EDM
Electric Discharge Machining.

FDM
Fused Deposition Modeling.

SLA
Stereo Lithography.

SLS
Selective Laser Sintering.

SLM
Selective Laser Melting.

J-P
Jetted Photopolymer (or Polyjet)

Newton's Law:

1. Every object remains stopped or goes on moving at a steady rate in a straight line unless acted upon by another force. This is the inertia principle.
2. The amount of force needed to make an object change its speed depends on the mass of the object and the amount of the acceleration or deceleration required.
3. To every action there is an equal and opposite reaction. When a body is pushed on way by a force, another force pushes back with equal strength.

<u>Polymers:</u>

A polymer is made of one or more large molecules formed from thousands of smaller molecules. Rubber and Wood are natural polymers. Plastics are synthetic (artificially made) polymers.

<u>Speed and Velocity:</u>

- Speed is the rate at which a moving object changes position (how far it moves in a fixed time).
- Velocity is speed in a particular direction.
- If either speed or direction is changed, velocity also changes.

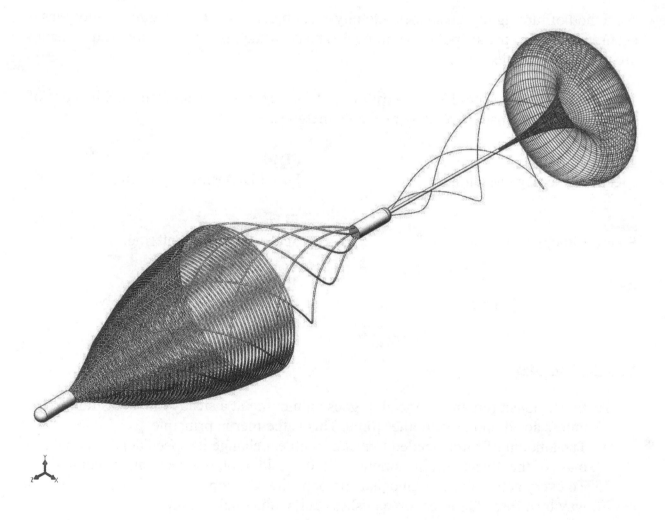

Absorbed

A feature, sketch, or annotation that is contained in another item (usually a feature) in the FeatureManager design tree. Examples are the profile sketch and profile path in a base-sweep, or a cosmetic thread annotation in a hole.

Align

Tools that assist in lining up annotations and dimensions (left, right, top, bottom, and so on). For aligning parts in an assembly.

Alternate position view

A drawing view in which one or more views are superimposed in phantom lines on the original view. Alternate position views are often used to show range of motion of an assembly.

Anchor point

The end of a leader that attaches to the note, block, or other annotation. Sheet formats contain anchor points for a bill of materials, a hole table, a revision table, and a weldment cut list.

Annotation

A text note or a symbol that adds specific design intent to a part, assembly, or drawing. Specific types of annotations include note, hole callout, surface finish symbol, datum feature symbol, datum target, geometric tolerance symbol, weld symbol, balloon, and stacked balloon. Annotations that apply only to drawings include center mark, annotation centerline, area hatch, and block.

Appearance callouts

Callouts that display the colors and textures of the face, feature, body, and part under the entity selected and are a shortcut to editing colors and textures.

Area hatch

A crosshatch pattern or fill applied to a selected face or to a closed sketch in a drawing.

Assembly

A document in which parts, features, and other assemblies (sub-assemblies) are mated together. The parts and sub-assemblies exist in documents separate from the assembly. For example, in an assembly, a piston can be mated to other parts, such as a connecting rod or cylinder. This new assembly can then be used as a sub-assembly in an assembly of an engine. The extension for a SOLIDWORKS assembly file name is .SLDASM.

Attachment point

The end of a leader that attaches to the model (to an edge, vertex, or face, for example) or to a drawing sheet.

Axis

A straight line that can be used to create model geometry, features, or patterns. An axis can be made in a number of different ways, including using the intersection of two planes.

Balloon

Labels parts in an assembly, typically including item numbers and quantity. In drawings, the item numbers are related to rows in a bill of materials.

Base

The first solid feature of a part.

Baseline dimensions

Sets of dimensions measured from the same edge or vertex in a drawing.

Bend

A feature in a sheet metal part. A bend generated from a filleted corner, cylindrical face, or conical face is a round bend; a bend generated from sketched straight lines is a sharp bend.

Bill of materials

A table inserted into a drawing to keep a record of the parts used in an assembly.

Block

A user-defined annotation that you can use in parts, assemblies, and drawings. A block can contain text, sketch entities (except points), and area hatch, and it can be saved in a file for later use as, for example, a custom callout or a company logo.

Bottom-up assembly

An assembly modeling technique where you create parts and then insert them into an assembly.

Broken-out section

A drawing view that exposes inner details of a drawing view by removing material from a closed profile, usually a spline.

Cavity

The mold half that holds the cavity feature of the design part.

Center mark

A cross that marks the center of a circle or arc.

Centerline

A centerline marks, in phantom font, an axis of symmetry in a sketch or drawing.

Chamfer

Bevels a selected edge or vertex. You can apply chamfers to both sketches and features.

Child

A dependent feature related to a previously-built feature. For example, a chamfer on the edge of a hole is a child of the parent hole.

Click-release

As you sketch, if you click and then release the pointer, you are in click-release mode. Move the pointer and click again to define the next point in the sketch sequence.

Click-drag

As you sketch, if you click and drag the pointer, you are in click-drag mode. When you release the pointer, the sketch entity is complete.

Closed profile

Also called a closed contour, it is a sketch or sketch entity with no exposed endpoints, for example, a circle or polygon.

Collapse

The opposite of explode. The collapse action returns an exploded assembly's parts to their normal positions.

Collision Detection

An assembly function that detects collisions between components when components move or rotate. A collision occurs when an entity on one component coincides with any entity on another component.

Component

Any part or sub-assembly within an assembly.

Configuration

A variation of a part or assembly within a single document. Variations can include different dimensions, features, and properties. For example, a single part such as a bolt can contain different configurations that vary the diameter and length.

ConfigurationManager

Located on the left side of the SOLIDWORKS window, it is a means to create, select, and view the configurations of parts and assemblies.

Constraint

The relations between sketch entities, or between sketch entities and planes, axes, edges, or vertices.

Construction geometry

The characteristic of a sketch entity that the entity is used in creating other geometry but is not itself used in creating features.

Coordinate system

A system of planes used to assign Cartesian coordinates to features, parts, and assemblies. Part and assembly documents contain default coordinate systems; other coordinate systems can be defined with reference geometry. Coordinate systems can be used with measurement tools and for exporting documents to other file formats.

Cosmetic thread

An annotation that represents threads.

Crosshatch

A pattern (or fill) applied to drawing views such as section views and broken-out sections.

Curvature

Curvature is equal to the inverse of the radius of the curve. The curvature can be displayed in different colors according to the local radius (usually of a surface).

Cut

A feature that removes material from a part by such actions as extrude, revolve, loft, sweep, thicken, cavity, and so on.

Dangling

A dimension, relation, or drawing section view that is unresolved. For example, if a piece of geometry is dimensioned, and that geometry is later deleted, the dimension becomes dangling.

Degrees of freedom

Geometry that is not defined by dimensions or relations is free to move. In 2D sketches, there are three degrees of freedom: movement along the X and Y axes, and rotation about the Z axis (the axis normal to the sketch plane). In 3D sketches and in assemblies, there are six degrees of freedom: movement along the X, Y, and Z axes, and rotation about the X, Y, and Z axes.

Derived part
A derived part is a new base, mirror, or component part created directly from an existing part and linked to the original part such that changes to the original part are reflected in the derived part.

Derived sketch
A copy of a sketch, in either the same part or the same assembly that is connected to the original sketch. Changes in the original sketch are reflected in the derived sketch.

Design Library
Located in the Task Pane, the Design Library provides a central location for reusable elements such as parts, assemblies, and so on.

Design table
An Excel spreadsheet that is used to create multiple configurations in a part or assembly document.

Detached drawing
A drawing format that allows opening and working in a drawing without loading the corresponding models into memory. The models are loaded on an as-needed basis.

Detail view
A portion of a larger view, usually at a larger scale than the original view.

Dimension line
A linear dimension line references the dimension text to extension lines indicating the entity being measured. An angular dimension line references the dimension text directly to the measured object.

DimXpertManager
Located on the left side of the SolidWorks window, it is a means to manage dimensions and tolerances created using DimXpert for parts according to the requirements of the ASME Y.14.41-2003 standard.

DisplayManager
The DisplayManager lists the appearances, decals, lights, scene, and cameras applied to the current model. From the DisplayManager, you can view applied content, and add, edit, or delete items. When PhotoView 360 is added in, the DisplayManager also provides access to PhotoView options.

Document
A file containing a part, assembly, or drawing.

Draft
The degree of taper or angle of a face, usually applied to molds or castings.

Drawing
A 2D representation of a 3D part or assembly. The extension for a SolidWorks drawing file name is .SLDDRW.

Drawing sheet
A page in a drawing document.

Driven dimension
Measurements of the model, but they do not drive the model and their values cannot be changed.

Driving dimension
Also referred to as a model dimension, it sets the value for a sketch entity. It can also control distance, thickness, and feature parameters.

Edge
A single outside boundary of a feature.

Edge flange
A sheet metal feature that combines a bend and a tab in a single operation.

Equation
Creates a mathematical relation between sketch dimensions, using dimension names as variables, or between feature parameters, such as the depth of an extruded feature or the instance count in a pattern.

Exploded view
Shows an assembly with its components separated from one another, usually to show how to assemble the mechanism.

Export
Save a SOLIDWORKS document in another format for use in other CAD/CAM, rapid prototyping, web, or graphics software applications.

Extension line
The line extending from the model indicating the point from which a dimension is measured.

Extrude

A feature that linearly projects a sketch to either add material to a part (in a base or boss) or remove material from a part (in a cut or hole).

Face

A selectable area (planar or otherwise) of a model or surface with boundaries that help define the shape of the model or surface. For example, a rectangular solid has six faces.

Fasteners

A SolidWorks Toolbox library that adds fasteners automatically to holes in an assembly.

Feature

An individual shape that, combined with other features, makes up a part or assembly. Some features, such as bosses and cuts, originate as sketches. Other features, such as shells and fillets, modify a feature's geometry. However, not all features have associated geometry. Features are always listed in the FeatureManager design tree.

FeatureManager design tree

Located on the left side of the SOLIDWORKS window, it provides an outline view of the active part, assembly, or drawing.

Fill

A solid area hatch or crosshatch. Fill also applies to patches on surfaces.

Fillet

An internal rounding of a corner or edge in a sketch, or an edge on a surface or solid.

Forming tool

Dies that bend, stretch, or otherwise form sheet metal to create such form features as louvers, lances, flanges, and ribs.

Fully defined

A sketch where all lines and curves in the sketch, and their positions, are described by dimensions or relations, or both, and cannot be moved. Fully defined sketch entities are shown in black.

Geometric tolerance

A set of standard symbols that specify the geometric characteristics and dimensional requirements of a feature.

Graphics area

The area in the SOLIDWORKS window where the part, assembly, or drawing appears.

Guide curve

A 2D or 3D curve used to guide a sweep or loft.

Handle

An arrow, square, or circle that you can drag to adjust the size or position of an entity (a feature, dimension, or sketch entity, for example).

Helix

A curve defined by pitch, revolutions, and height. A helix can be used, for example, as a path for a swept feature cutting threads in a bolt.

Hem

A sheet metal feature that folds back at the edge of a part. A hem can be open, closed, double, or tear-drop.

HLR

(Hidden lines removed) a view mode in which all edges of the model that are not visible from the current view angle are removed from the display.

HLV

(Hidden lines visible) A view mode in which all edges of the model that are not visible from the current view angle are shown gray or dashed.

Import

Open files from other CAD software applications into a SOLIDWORKS document.

In-context feature

A feature with an external reference to the geometry of another component; the in-context feature changes automatically if the geometry of the referenced model or feature changes.

Inference

The system automatically creates (infers) relations between dragged entities (sketched entities, annotations, and components) and other entities and geometry. This is useful when positioning entities relative to one another.

Instance

An item in a pattern or a component in an assembly that occurs more than once. Blocks are inserted into drawings as instances of block definitions.

Interference detection
A tool that displays any interference between selected components in an assembly.

Jog
A sheet metal feature that adds material to a part by creating two bends from a sketched line.

Knit
A tool that combines two or more faces or surfaces into one. The edges of the surfaces must be adjacent and not overlapping, but they cannot ever be planar. There is no difference in the appearance of the face or the surface after knitting.

Layout sketch
A sketch that contains important sketch entities, dimensions, and relations. You reference the entities in the layout sketch when creating new sketches, building new geometry, or positioning components in an assembly. This allows for easier updating of your model because changes you make to the layout sketch propagate to the entire model.

Leader
A solid line from an annotation (note, dimension, and so on) to the referenced feature.

Library feature
A frequently used feature, or combination of features, that is created once and then saved for future use.

Lightweight
A part in an assembly or a drawing has only a subset of its model data loaded into memory. The remaining model data is loaded on an as-needed basis. This improves performance of large and complex assemblies.

Line
A straight sketch entity with two endpoints. A line can be created by projecting an external entity such as an edge, plane, axis, or sketch curve into the sketch.

Loft
A base, boss, cut, or surface feature created by transitions between profiles.

Lofted bend
A sheet metal feature that produces a roll form or a transitional shape from two open profile sketches. Lofted bends often create funnels and chutes.

Mass properties
A tool that evaluates the characteristics of a part or an assembly such as volume, surface area, centroid, and so on.

Mate
A geometric relationship, such as coincident, perpendicular, tangent, and so on, between parts in an assembly.

Mate reference
Specifies one or more entities of a component to use for automatic mating. When you drag a component with a mate reference into an assembly, the software tries to find other combinations of the same mate reference name and mate type.

Mates folder
A collection of mates that are solved together. The order in which the mates appear within the Mates folder does not matter.

Mirror
(a) A mirror feature is a copy of a selected feature, mirrored about a plane or planar face.
(b) A mirror sketch entity is a copy of a selected sketch entity that is mirrored about a centerline.

Miter flange
A sheet metal feature that joins multiple edge flanges together and miters the corner.

Model
3D solid geometry in a part or assembly document. If a part or assembly document contains multiple configurations, each configuration is a separate model.

Model dimension
A dimension specified in a sketch or a feature in a part or assembly document that defines some entity in a 3D model.

Model item
A characteristic or dimension of feature geometry that can be used in detailing drawings.

Model view
A drawing view of a part or assembly.

Mold

A set of manufacturing tooling used to shape molten plastic or other material into a designed part. You design the mold using a sequence of integrated tools that result in cavity and core blocks that are derived parts of the part to be molded.

Motion Study

Motion Studies are graphical simulations of motion and visual properties with assembly models. Analogous to a configuration, they do not actually change the original assembly model or its properties. They display the model as it changes based on simulation elements you add.

Multibody part

A part with separate solid bodies within the same part document. Unlike the components in an assembly, multibody parts are not dynamic.

Native format

DXF and DWG files remain in their original format (are not converted into SOLIDWORKS format) when viewed in SOLIDWORKS drawing sheets (view only).

Open profile

Also called an open contour, it is a sketch or sketch entity with endpoints exposed. For example, a U-shaped profile is open.

Ordinate dimensions

A chain of dimensions measured from a zero ordinate in a drawing or sketch.

Origin

The model origin appears as three gray arrows and represents the (0,0,0) coordinate of the model. When a sketch is active, a sketch origin appears in red and represents the (0,0,0) coordinate of the sketch. Dimensions and relations can be added to the model origin, but not to a sketch origin.

Out-of-context feature

A feature with an external reference to the geometry of another component that is not open.

Over defined

A sketch is over defined when dimensions or relations are either in conflict or redundant.

Parameter

A value used to define a sketch or feature (often a dimension).

Parent

An existing feature upon which other features depend. For example, in a block with a hole, the block is the parent to the child hole feature.

Part

A single 3D object made up of features. A part can become a component in an assembly, and it can be represented in 2D in a drawing. Examples of parts are bolt, pin, plate, and so on. The extension for a SOLIDWORKS part file name is .SLDPRT.

Path

A sketch, edge, or curve used in creating a sweep or loft.

Pattern

A pattern repeats selected sketch entities, features, or components in an array, which can be linear, circular, or sketch-driven. If the seed entity is changed, the other instances in the pattern update.

Physical Dynamics

An assembly tool that displays the motion of assembly components in a realistic way. When you drag a component, the component applies a force to other components it touches. Components move only within their degrees of freedom.

Pierce relation

Makes a sketch point coincident to the location at which an axis, edge, line, or spline pierces the sketch plane.

Planar

Entities that can lie on one plane. For example, a circle is planar, but a helix is not.

Plane

Flat construction geometry. Planes can be used for a 2D sketch, section view of a model, a neutral plane in a draft feature, and others.

Point

A singular location in a sketch, or a projection into a sketch at a single location of an external entity (origin, vertex, axis, or point in an external sketch).

Predefined view

A drawing view in which the view position, orientation, and so on can be specified before a model is inserted. You can save drawing documents with predefined views as templates.

Profile

A sketch entity used to create a feature (such as a loft) or a drawing view (such as a detail view). A profile can be open (such as a U shape or open spline) or closed (such as a circle or closed spline).

Projected dimension

If you dimension entities in an isometric view, projected dimensions are the flat dimensions in 2D.

Projected view

A drawing view projected orthogonally from an existing view.

PropertyManager

Located on the left side of the SOLIDWORKS window, it is used for dynamic editing of sketch entities and most features.

RealView graphics

A hardware (graphics card) support of advanced shading in real time; the rendering applies to the model and is retained as you move or rotate a part.

Rebuild

Tool that updates (or regenerates) the document with any changes made since the last time the model was rebuilt. Rebuild is typically used after changing a model dimension.

Reference dimension

A dimension in a drawing that shows the measurement of an item but cannot drive the model and its value cannot be modified. When model dimensions change, reference dimensions update.

Reference geometry

Includes planes, axes, coordinate systems, and 3D curves. Reference geometry is used to assist in creating features such lofts, sweeps, drafts, chamfers, and patterns.

Relation

A geometric constraint between sketch entities or between a sketch entity and a plane, axis, edge, or vertex. Relations can be added automatically or manually.

Relative view

A relative (or relative to model) drawing view is created relative to planar surfaces in a part or assembly.

Reload
Refreshes shared documents. For example, if you open a part file for read-only access while another user makes changes to the same part, you can reload the new version, including the changes.

Reorder
Reordering (changing the order of) items is possible in the FeatureManager design tree. In parts, you can change the order in which features are solved. In assemblies, you can control the order in which components appear in a bill of materials.

Replace
Substitutes one or more open instances of a component in an assembly with a different component.

Resolved
A state of an assembly component (in an assembly or drawing document) in which it is fully loaded in memory. All the component's model data is available, so its entities can be selected, referenced, edited, and used in mates, and so on.

Revolve
A feature that creates a base or boss, a revolved cut, or revolved surface by revolving one or more sketched profiles around a centerline.

Rip
A sheet metal feature that removes material at an edge to allow a bend.

Rollback
Suppresses all items below the rollback bar.

Section
Another term for profile in sweeps.

Section line
A line or centerline sketched in a drawing view to create a section view.

Section scope
Specifies the components to be left uncut when you create an assembly drawing section view.

Section view
A section view (or section cut) is (1) a part or assembly view cut by a plane, or (2) a drawing view created by cutting another drawing view with a section line.

Seed

A sketch or an entity (a feature, face, or body) that is the basis for a pattern. If you edit the seed, the other entities in the pattern are updated.

Shaded

Displays a model as a colored solid.

Shared values

Also called linked values, these are named variables that you assign to set the value of two or more dimensions to be equal.

Sheet format

Includes page size and orientation, standard text, borders, title blocks, and so on. Sheet formats can be customized and saved for future use. Each sheet of a drawing document can have a different format.

Shell

A feature that hollows out a part, leaving open the selected faces and thin walls on the remaining faces. A hollow part is created when no faces are selected to be open.

Sketch

A collection of lines and other 2D objects on a plane or face that forms the basis for a feature such as a base or a boss. A 3D sketch is non-planar and can be used to guide a sweep or loft, for example.

Smart Fasteners

Automatically adds fasteners (bolts and screws) to an assembly using the SOLIDWORKS Toolbox library of fasteners.

SmartMates

An assembly mating relation that is created automatically.

Solid sweep

A cut sweep created by moving a tool body along a path to cut out 3D material from a model.

Spiral

A flat or 2D helix, defined by a circle, pitch, and number of revolutions.

Spline

A sketched 2D or 3D curve defined by a set of control points.

Split line

Projects a sketched curve onto a selected model face, dividing the face into multiple faces so that each can be selected individually. A split line can be used to create draft features, to create face blend fillets, and to radiate surfaces to cut molds.

Stacked balloon

A set of balloons with only one leader. The balloons can be stacked vertically (up or down) or horizontally (left or right).

Standard 3 views

The three orthographic views (front, right, and top) that are often the basis of a drawing.

StereoLithography

The process of creating rapid prototype parts using a faceted mesh representation in STL files.

Sub-assembly

An assembly document that is part of a larger assembly. For example, the steering mechanism of a car is a sub-assembly of the car.

Suppress

Removes an entity from the display and from any calculations in which it is involved. You can suppress features, assembly components, and so on. Suppressing an entity does not delete the entity; you can unsuppress the entity to restore it.

Surface

A zero-thickness planar or 3D entity with edge boundaries. Surfaces are often used to create solid features. Reference surfaces can be used to modify solid features.

Sweep

Creates a base, boss, cut, or surface feature by moving a profile (section) along a path. For cut-sweeps, you can create solid sweeps by moving a tool body along a path.

Tangent arc

An arc that is tangent to another entity, such as a line.

Tangent edge

The transition edge between rounded or filleted faces in hidden lines visible or hidden lines removed modes in drawings.

Task Pane

Located on the right-side of the SOLIDWORKS window, the Task Pane contains SOLIDWORKS Resources, the Design Library, and the File Explorer.

Template

A document (part, assembly, or drawing) that forms the basis of a new document. It can include user-defined parameters, annotations, predefined views, geometry, and so on.

Temporary axis

An axis created implicitly for every conical or cylindrical face in a model.

Thin feature

An extruded or revolved feature with constant wall thickness. Sheet metal parts are typically created from thin features.

TolAnalyst

A tolerance analysis application that determines the effects that dimensions and tolerances have on parts and assemblies.

Top-down design

An assembly modeling technique where you create parts in the context of an assembly by referencing the geometry of other components. Changes to the referenced components propagate to the parts that you create in context.

Triad

Three axes with arrows defining the X, Y, and Z directions. A reference triad appears in part and assembly documents to assist in orienting the viewing of models. Triads also assist when moving or rotating components in assemblies.

Under defined

A sketch is under defined when there are not enough dimensions and relations to prevent entities from moving or changing size.

Vertex

A point at which two or more lines or edges intersect. Vertices can be selected for sketching, dimensioning, and many other operations.

Viewports

Windows that display views of models. You can specify one, two, or four viewports. Viewports with orthogonal views can be linked, which links orientation and rotation.

Virtual sharp

A sketch point at the intersection of two entities after the intersection itself has been removed by a feature such as a fillet or chamfer. Dimensions and relations to the virtual sharp are retained even though the actual intersection no longer exists.

Weldment
A multibody part with structural members.

Weldment cut list
A table that tabulates the bodies in a weldment along with descriptions and lengths.

Wireframe
A view mode in which all edges of the part or assembly are displayed.

Zebra stripes
Simulate the reflection of long strips of light on a very shiny surface. They allow you to see small changes in a surface that may be hard to see with a standard display.

Zoom
To simulate movement toward or away from a part or an assembly.

Learn Mold-Tooling Designs with SOLIDWORKS Advanced Techniques Textbook

Index

3

3d path, 1-15

3d sketch, 1-1, 1-5, 1-11, 1-17, 1-18, 1-20, 11-18

3-point Arc, 4-5, 7-3, 8-3, 8-4, 8-5, 8-32, 11-31, 20-9, 20-21

A

add a curve, 10-29

add ins, 16-15

add points, 10-29

advanced loft, 11-15

advanced sweep, 11-16

all bodies, 15-28

along z relation, 16-11

anchor, 12-5

anisotropic, 12-18, 12-19

ansi inch, 16-12

appearances, 3-15, 9-24, 23-1

apply to all edges, 8-26

arc conditions, 3-3

arc slot, 14-5, 14-8

assembly features, 16-11

at angle, 1-10, 1-11, 2-7, 4-4

auto relief, 13-21, 15-5, 16-8

auto rotate view, 21-4

auto spacing, 15-17

avi, 16-17

axis indicator, 1-3

axis, 11-18, 13-38, 13-39

B

base feature, 3-4, 13-3, 13-17, 13-21, 13-40, 14-15, 14-32

base flange, 13-3

begin assembly, 20-3

bend allowance, 13-1, 13-18, 15-13, 16-8

bend table, 13-1, 13-17

bend table, 15-13

bi-directional, 14-6

blind extrude, 5-11

boundary surface, 9-17, 10-9

break all, 22-3, 22-9

bridge lance, 14-23, 14-24, 14-25, 14-26

C

cad formats, 15-1

cap ends, 14-6

cavity, 18-1, 18-2, 18-6, 18-8, 18-10, 18-14

center rectangle, 13-25, 14-3, 18-7

centerline parameters, 6-11, 8-23

centerline, 20-5, 20-6, 20-23, 19-33

centroid, 19-25

chamfers, 6-18, 14-30, 21-8, 21-9, 21-19

circle, 5-9

circular component pattern, 15-29

circular pattern, 2-15, 6-22, 11-20, 13-38, 13-39

circular sketch pattern, 4-27

clear mesh faces, 5-10

close curve, 7-6, 7-17, 7-35

close loft, 7-10

closed curve, 4-18

coincident mate, 16-4

coincident relation, 4-26

coincident, 2-9

collapse, 19-36

collect all bends, 14-18, 14-20, 14-22, 15-16

combine common, 15-28

components, 16-4, 20-1, 20-30

composite curve, 1-17, 1-20, 1-21, 4-1, 4-2, 4-8, 4-9, 4-11, 4-25, 4-27, 11-5

concentric relation, 3-8, 20-17, 20-26

configuration, 15-20, 15-21

connector, 6-4

connectors, 5-10, 6-11

constant fillet, 9-23

construction geometry, 3-8, 3-19, 3-20

construction, 7-3, 7-4, 7-11, 7-12

Designed and Rendered with SOLIDWORKS

SOLIDWORKS Quick Guide

STANDARD Toolbar

 Creates a new document.

 Opens an existing document.

 Saves an active document.

 Make Drawing from Part/Assembly.

 Make Assembly from Part/Assembly.

 Prints the active document.

 Print preview.

 Cuts the selection & puts it on the clipboard.

 Copies the selection & puts it on the clipboard.

 Inserts the clipboard contents.

 Deletes the selection.

 Reverses the last action.

 Rebuilds the part / assembly / drawing.

 Redo the last action that was undone.

 Saves all documents.

 Edits material.

 Closes an existing document.

 Shows or hides the Selection Filter toolbar.

 Shows or hides the Web toolbar.

 Properties.

 File properties.

STANDARD (right column)

 Loads or unloads the 3D instant website add-in.

 Select tool.

 Select the entire document.

 Checks read-only files.

 Options.

 Help.

 Full screen view.

 OK.

 Cancel.

 Magnified selection.

SKETCH TOOLS Toolbar

 Select.

 Sketch.

 3D Sketch.

 Sketches a rectangle from the center.

 Sketches a centerpoint arc slot.

 Sketches a 3-point arc slot.

 Sketches a straight slot.

 Sketches a centerpoint straight slot.

 Sketches a 3-point arc.

 Creates sketched ellipses.

Quick Reference Guide to SOLIDWORKS Command Icons & Toolbars

SKETCH TOOLS Toolbar

3D sketch on plane.

Sets up Grid parameters.

Creates a sketch on a selected plane or face.

Equation driven curve.

Modifies a sketch.

Copies sketch entities.

Scales sketch entities.

Rotates sketch entities.

Sketches a 3 point rectangle from the center.

Sketches 3 point corner rectangle.

Sketches a line.

Creates a center point arc: center, start, end.

Creates an arc tangent to a line.

Sketches splines on a surface or face.

Sketches a circle.

Sketches a circle by its perimeter.

Makes a path of sketch entities.

Mirrors entities dynamically about a centerline.

Insert a plane into the 3D sketch.

Instant 2D.

Sketch numeric input.

Detaches segment on drag.

Sketch picture.

Partial ellipses.

Adds a Parabola.

Adds a spline.

Sketches a polygon.

Sketches a corner rectangle.

Sketches a parallelogram.

Creates points.

Creates sketched centerlines.

Adds text to sketch.

Converts selected model edges or sketch entities to sketch segments.

Creates a sketch along the intersection of multiple bodies.

Converts face curves on the selected face into 3D sketch entities.

Mirrors selected segments about a centerline.

Fillets the corner of two lines.

Creates a chamfer between two sketch entities.

Creates a sketch curve by offsetting model edges or sketch entities at a specified distance.

Trims a sketch segment.

Extends a sketch segment.

Splits a sketch segment.

Construction Geometry.

Creates linear steps and repeat of sketch entities.

Creates circular steps and repeat of sketch entities.

Quick Reference Guide to SOLIDWORKS Command Icons & Toolbars

SHEET METAL Toolbar

 Add a bend from a selected sketch in a Sheet Metal part.

 Shows flat pattern for this sheet metal part.

 Shows part without inserting any bends.

 Inserts a rip feature to a sheet metal part.

 Create a Sheet Metal part or add material to existing Sheet Metal part.

 Inserts a Sheet Metal Miter Flange feature.

 Folds selected bends.

 Unfolds selected bends.

 Inserts bends using a sketch line.

 Inserts a flange by pulling an edge.

 Inserts a sheet metal corner feature.

 Inserts a Hem feature by selecting edges.

 Breaks a corner by filleting/chamfering it.

 Inserts a Jog feature using a sketch line.

 Inserts a lofted bend feature using 2 sketches.

 Creates inverse dent on a sheet metal part.

 Trims out material from a corner, in a sheet metal part.

 Inserts a fillet weld bead.

 Converts a solid/surface into a sheet metal part.

 Adds a Cross Break feature into a selected face.

 Sweeps an open profile along an open/closed path.

 Adds a gusset/rib across a bend.

 Corner relief.

 Welds the selected corner.

SURFACES Toolbar

 Creates mid surfaces between offset face pairs.

 Patches surface holes and external edges.

 Creates an extruded surface.

 Creates a revolved surface.

 Creates a swept surface.

 Creates a lofted surface.

 Creates an offset surface.

 Radiates a surface originating from a curve, parallel to a plane.

 Knits surfaces together.

 Creates a planar surface from a sketch or a set of edges.

 Creates a surface by importing data from a file.

 Extends a surface.

 Trims a surface.

 Surface flatten.

 Deletes Face(s).

 Replaces Face with Surface.

 Patches surface holes and external edges by extending the surfaces.

 Creates parting surfaces between core & cavity surfaces.

 Inserts ruled surfaces from edges.

WELDMENTS Toolbar

 Creates a weldment feature.

 Creates a structure member feature.

 Adds a gusset feature between 2 planar adjoining faces.

 Creates an end cap feature.

 Adds a fillet weld bead feature.

 Trims or extends structure members.

 Weld bead.

Quick Reference Guide to SOLIDWORKS Command Icons & Toolbars

DIMENSIONS/RELATIONS Toolbar

 Inserts dimension between two lines.

 Creates a horizontal dimension between selected entities.

 Creates a vertical dimension between selected entities.

 Creates a reference dimension between selected entities.

 Creates a set of ordinate dimensions.

 Creates a set of Horizontal ordinate dimensions.

 Creates a set of Vertical ordinate dimensions.

 Creates a chamfer dimension.

 Adds a geometric relation.

 Automatically Adds Dimensions to the current sketch.

 Displays and deletes geometric relations.

 Fully defines a sketch.

 Scans a sketch for elements of equal length or radius.

 Angular Running dimension.

 Display / Delete dimension.

 Isolate changed dimension.

 Path length dimension.

BLOCK Toolbar

 Makes a new block.

 Edits the selected block.

 Inserts a new block to a sketch or drawing.

Adds/Removes sketch entities to/from blocks.

 Updates parent sketches affected by this block.

 Saves the block to a file.

 Explodes the selected block.

 Inserts a belt.

STANDARD VIEWS Toolbar

 Front view.

 Back view.

 Left view.

 Right view.

 Top view.

 Bottom view.

 Isometric view.

 Trimetric view.

 Dimetric view.

 Normal to view.

 Links all views in the viewport together.

 Displays viewport with front & right views.

 Displays a 4 view viewport with 1st or 3rd angle of projection.

 Displays viewport with front & top.

 Displays viewport with a single view.

 View selector.

New view.

FEATURES Toolbar

 Creates a boss feature by extruding a sketched profile.

 Creates a revolved feature based on profile and angle parameter.

 Creates a cut feature by extruding a sketched profile.

 Creates a cut feature by revolving a sketched profile.

 Thread.

 Creates a cut by sweeping a closed profile along an open or closed path.

 Loft cut.

 Creates a cut by thickening one or more adjacent surfaces.

 Adds a deformed surface by push or pull on points.

 Creates a lofted feature between two or more profiles.

 Creates a solid feature by thickening one or more adjacent surfaces.

 Creates a filled feature.

 Chamfers an edge or a chain of tangent edges.

 Inserts a rib feature.

 Combine.

 Creates a shell feature.

 Applies draft to a selected surface.

 Creates a cylindrical hole.

 Inserts a hole with a pre-defined cross section.

 Puts a dome surface on a face.

 Model break view.

 Applies global deformation to solid or surface bodies.

 Wraps closed sketch contour(s) onto a face.

 Curve Driven pattern.

 Suppresses the selected feature or component.

 Un-suppresses the selected feature or component.

 Flexes solid and surface bodies.

 Intersect.

 Variable Patterns.

 Live Section Plane.

 Mirrors.

 Scale.

 Creates a Sketch Driven pattern.

 Creates a Table Driven Pattern.

 Inserts a split Feature.

 Hole series.

 Joins bodies from one or more parts into a single part in the context of an assembly.

 Deletes a solid or a surface.

 Instant 3D.

 Inserts a part from file into the active part document.

 Moves/Copies solid and surface bodies or moves graphics bodies.

 Merges short edges on faces.

 Pushes solid / surface model by another solid / surface model.

 Moves face(s) of a solid.

 FeatureWorks Options.

 Linear Pattern.

 Fill Pattern.

 Cuts a solid model with a surface.

 Boundary Boss/Base.

 Boundary Cut.

 Circular Pattern.

 Recognize Features.

 Grid System.

MOLD TOOLS Toolbar

 Extracts core(s) from existing tooling split.

 Constructs a surface patch.

 Moves face(s) of a solid.

 Creates offset surfaces.

 Inserts cavity into a base part.

 Scales a model by a specified factor.

 Applies draft to a selected surface.

 Inserts a split line feature.

 Creates parting lines to separate core & cavity surfaces.

 Finds & creates mold shut-off surfaces.

 Creates a planar surface from a sketch or a set of edges.

 Knits surfaces together.

 Inserts ruled surfaces from edges.

 Creates parting surfaces between core & cavity surfaces.

 Creates multiple bodies from a single body.

 Inserts a tooling split feature.

 Creates parting surfaces between the core & cavity.

 Inserts surface body folders for mold operation.

SELECTION FILTERS Toolbar

 Turns selection filters on and off.

 Clears all filters.

 Selects all filters.

 Inverts current selection.

 Allows selection of edges only.

 Allows selection filter for vertices only.

 Allows selection of faces only.

 Adds filter for Surface Bodies.

 Adds filter for Solid Bodies.

 Adds filter for Axes.

 Adds filter for Planes.

 Adds filter for Sketch Points.

 Allows selection for sketch only.

 Adds filter for Sketch Segments.

 Adds filter for Midpoints.

 Adds filter for Center Marks.

 Adds filter for Centerline.

 Adds filter for Dimensions and Hole Callouts.

 Adds filter for Surface Finish Symbols.

 Adds filter for Geometric Tolerances.

 Adds filter for Notes / Balloons.

 Adds filter for Weld Symbols.

 Adds filter for Weld beads.

 Adds filter for Datum Targets.

 Adds filter for Datum feature only.

 Adds filter for blocks.

 Adds filter for Cosmetic Threads.

Adds filter for Dowel pin symbols.

 Adds filter for connection points.

 Adds filter for routing points.

SOLIDWORKS Add-Ins Toolbar

 Loads/unloads CircuitWorks add-in.

 Loads/unloads the Design Checker add-in.

 Loads/unloads the PhotoView 360 add-in.

 Loads/unloads the Scan-to-3D add-in.

 Loads/unloads the SOLIDWORKS Motions add-in.

 Loads/unloads the SOLIDWORKS Routing add-in.

 Loads/unloads the SOLIDWORKS Simulation add-in.

 Loads/unloads the SOLIDWORKS Toolbox add-in.

 Loads/unloads the SOLIDWORKS TolAnalysis add-in.

 Loads/unloads the SOLIDWORKS Flow Simulation add-in.

 Loads/unloads the SOLIDWORKS Plastics add-in.

 Loads/unloads the SOLIDWORKS MBD SNL license.

FASTENING FEATURES Toolbar

 Creates a parameterized mounting boss.

 Creates a parameterized snap hook.

 Creates a groove to mate with a hook feature.

 Uses sketch elements to create a vent for air flow.

 Creates a lip/groove feature.

SCREEN CAPTURE Toolbar

 Copies the current graphics window to the clipboard.

 Records the current graphics window to an AVI file.

 Stops recording the current graphics window to an AVI file.

EXPLODE LINE SKETCH Toolbar

 Adds a route line that connect entities.

 Adds a jog to the route lines.

LINE FORMAT Toolbar

 Changes layer properties.

 Changes the current document layer.

 Changes line color.

 Changes line thickness.

 Changes line style.

 Hides / Shows a hidden edge.

 Changes line display mode.

Did you know??

* Ctrl+Q will force a rebuild on all features of a part.

* Ctrl+B will rebuild the feature being worked on and its dependents.

2D-To-3D Toolbar

 Makes a Front sketch from the selected entities.

 Makes a Top sketch from the selected entities.

 Makes a Right sketch from the selected entities.

Makes a Left sketch from the selected entities.

Makes a Bottom sketch from the selected entities.

Makes a Back sketch from the selected entities.

Makes an Auxiliary sketch from the selected entities.

Creates a new sketch from the selected entities.

Repairs the selected sketch.

Aligns a sketch to the selected point.

Creates an extrusion from the selected sketch segments, starting at the selected sketch point.

Creates a cut from the selected sketch segments, optionally starting at the selected sketch point.

ALIGN Toolbar

Aligns the left side of the selected annotations with the leftmost annotation.

Aligns the right side of the selected annotations with the rightmost annotation.

Aligns the top side of the selected annotations with the topmost annotation.

Aligns the bottom side of the selected annotations with the lowermost annotation.

Evenly spaces the selected annotations horizontally.

Evenly spaces the selected annotations vertically.

Centrally aligns the selected annotations horizontally.

Centrally aligns the selected annotations vertically.

Compacts the selected annotations horizontally.

Compacts the selected annotations vertically.

Creates a group from the selected items.

Deletes the grouping between these items.

Aligns & groups selected dimensions along a line or an arc.

Aligns & groups dimensions at uniform distances.

Evenly spaces selected dimensions.

Aligns collinear selected dimensions.

Aligns stagger selected dimensions.

SOLIDWORKS MBD Toolbar

Captures 3D view.

Manages 3D PDF templates.

Creates shareable 3D PDF presentations.

Toggles dynamic annotation views.

MACRO Toolbar

Runs a Macro.

Stops Macro recorder.

Records (or pauses recording of) actions to create a Macro.

Launches the Macro Editor and begins editing a new macro.

Opens a Macro file for editing.

Creates a custom macro.

SMARTMATES icons

Concentric & Coincident 2 circular edges.

Concentric 2 cylindrical faces.

Coincident 2 linear edges.

Coincident 2 planar faces.

Coincident 2 vertices.

Coincident 2 origins or coordinate systems.

TABLE Toolbar

 Adds a hole table of selected holes from a specified origin datum.

 Adds a Bill of Materials.

 Adds a revision table.

 Displays a Design table in a drawing.

 Adds a weldments cuts list table.

 Adds a Excel based of Bill of Materials

 Adds a weldment cut list table.

REFERENCE GEOMETRY Toolbar

 Adds a reference plane.

 Creates an axis.

 Creates a coordinate system.

 Adds the center of mass.

 Specifies entities to use as references using SmartMates.

SPLINE TOOLS Toolbar

 Inserts a point to a spline.

 Displays all points where the concavity of selected spline changes.

 Displays minimum radius of selected spline.

 Displays curvature combs of selected spline.

 Reduces numbers of points in a selected spline.

 Adds a tangency control.

 Adds a curvature control.

 Adds a spline based on selected sketch entities & edges.

 Displays the spline control polygon.

ANNOTATIONS Toolbar

 Inserts a note.

 Inserts a surface finish symbol.

 Inserts a new geometric tolerancing symbol.

 Attaches a balloon to the selected edge or face.

 Adds balloons for all components in selected view.

 Inserts a stacked balloon.

 Attaches a datum feature symbol to a selected edge / detail.

 Inserts a weld symbol on the selected edge / face / vertex.

 Inserts a datum target symbol and / or point attached to a selected edge / line.

 Selects and inserts block.

 Inserts annotations & reference geometry from the part / assembly into the selected.

 Adds center marks to circles on model.

 Inserts a Centerline.

 Inserts a hole callout.

 Adds a cosmetic thread to the selected cylindrical feature.

 Inserts a Multi-Jog leader.

 Selects a circular edge or and arc for Dowel pin symbol insertion.

 Adds a view location symbol.

 Inserts latest version symbol.

 Adds a cross hatch patterns or solid fill.

 Adds a weld bead caterpillar on an edge.

 Adds a weld symbol on a selected entity.

 Inserts a revision cloud.

 Inserts a magnetic line.

 Hides/shows annotation.

DRAWINGS Toolbar

 Updates the selected view to the model's current stage.

 Creates a detail view.

 Creates a section view.

 Inserts an Alternate Position view.

 Unfolds a new view from an existing view.

 Generates a standard 3-view drawing (1st or 3rd angle).

 Inserts an auxiliary view of an inclined surface.

 Adds an Orthogonal or Named view based on an existing part or assembly.

 Adds a Relative view by two orthogonal faces or planes.

 Adds a Predefined orthogonal projected or Named view with a model.

 Adds an empty view.

 Adds vertical break lines to selected view.

 Crops a view.

 Creates a Broken-out section.

QUICK SNAP Toolbar

 Snap to points.

 Snap to center points.

 Snap to midpoints.

 Snap to quadrant points.

 Snap to intersection of 2 curves.

 Snap to nearest curve.

 Snap tangent to curve.

 Snap perpendicular to curve.

Snap parallel to line.

 Snap horizontally / vertically to points.

 Snap horizontally / vertically.

 Snap to discrete line lengths.

 Snap to angle.

LAYOUT Toolbar

 Creates the assembly layout sketch.

 Sketches a line.

 Sketches a corner rectangle.

 Sketches a circle.

 Sketches a 3 point arc.

 Rounds a corner.

 Trims or extends a sketch.

 Adds sketch entities by offsetting faces, edges curves.

 Mirrors selected entities about a centerline.

 Adds a relation.

 Creates a dimension.

 Displays / Deletes geometric relations.

 Makes a new block.

 Edits the selected block.

 Inserts a new block to the sketch or drawing.

 Adds / Removes sketch entities to / from a block.

 Saves the block to a file.

 Explodes the selected block.

 Creates a new part from a layout sketch block.

 Positions 2 components relative to one another.

CURVES Toolbar

 Projects sketch onto selected surface.

 Inserts a split line feature.

 Creates a composite curve from selected edges, curves and sketches.

 Creates a curve through free points.

 Creates a 3D curve through reference points.

 Helical curve defined by a base sketch and shape parameters.

VIEW Toolbar

 Displays a view in the selected orientation.

 Reverts to previous view.

 Redraws the current window.

 Zooms out to see entire model.

 Zooms in by dragging a bounding box.

 Zooms in or out by dragging up or down.

 Zooms to fit all selected entities.

 Dynamic view rotation.

 Scrolls view by dragging.

 Displays image in wireframe mode.

 Displays hidden edges in gray.

 Displays image with hidden lines removed.

 Controls the visibility of planes.

 Controls the visibility of axis.

 Controls the visibility of parting lines.

 Controls the visibility of temporary axis.

 Controls the visibility of origins.

 Controls the visibility of coordinate systems.

 Controls the visibility of reference curves.

 Controls the visibility of sketches.

 Controls the visibility of 3D sketch planes.

 Controls the visibility of 3D sketch.

 Controls the visibility of all annotations.

 Controls the visibility of reference points.

 Controls the visibility of routing points.

 Controls the visibility of lights.

 Controls the visibility of cameras.

 Controls the visibility of sketch relations.

 Changes the display state for the current configuration.

 Rolls the model view.

 Turns the orientation of the model view.

 Dynamically manipulate the model view in 3D to make selection.

 Changes the display style for the active view.

 Displays a shade view of the model with its edges.

 Displays a shade view of the model.

 Toggles between draft quality & high quality HLV.

 Cycles through or applies a specific scene.

 Views the models through one of the model's cameras.

 Displays a part or assembly w/different colors according to the local radius of curvature.

 Displays zebra stripes.

 Displays a model with hardware accelerated shades.

 Applies a cartoon affect to model edges & faces.

 Views simulations symbols.

TOOLS Toolbar

 Calculates the distance between selected items.

 Adds or edits equation.

 Calculates the mass properties of the model.

 Checks the model for geometry errors.

 Inserts or edits a Design Table.

 Evaluates section properties for faces and sketches that lie in parallel planes.

 Reports Statistics for this Part/Assembly.

 Deviation Analysis.

 Runs the SimulationXpress analysis wizard powered by SOLIDWORKS Simulation.

 Checks the spelling.

 Import diagnostics.

 Runs the DFMXpress analysis wizard.

 Runs the SOLIDWORKSFloXpress analysis wizard.

ASSEMBLY Toolbar

 Creates a new part & inserts it into the assembly.

 Adds an existing part or sub-assembly to the assembly.

 Creates a new assembly & inserts it into the assembly.

 Turns on/off large assembly mode for this document.

 Hides / shows model(s) associated with the selected model(s).

 Toggles the transparency of components.

 Changes the selected components to suppressed or resolved.

 Inserts a belt.

 Toggles between editing part and assembly.

 Smart Fasteners.

 Positions two components relative to one another.

 External references will not be created.

 Moves a component.

 Rotates an un-mated component around its center point.

 Replaces selected components.

 Replaces mate entities of mates of the selected components on the selected Mates group.

 Creates a New Exploded view.

 Creates or edits explode line sketch.

 Interference detection.

 Shows or Hides the Simulation toolbar.

 Patterns components in one or two linear directions.

 Patterns components around an axis.

 Sets the transparency of the components other than the one being edited.

 Sketch driven component pattern.

 Pattern driven component pattern.

 Curve driven component pattern.

 Chain driven component pattern.

 SmartMates by dragging & dropping components.

 Checks assembly hole alignments.

 Mirrors subassemblies and parts.

To add or remove an icon
to or from the toolbar, first select:

Tools/Customize/Commands

Next, select a **Category**, click a button to see its description and then drag/drop the command icon into any toolbar.

SOLIDWORKS Quick-Guide©
STANDARD Keyboard Shortcuts

Rotate the model

* Horizontally or Vertically: _____ Arrow keys

* Horizontally or Vertically 90°: _____ Shift + Arrow keys

* Clockwise or Counterclockwise: _____ Alt + left or right Arrow

* Pan the model: _____ Ctrl + Arrow keys

* Zoom in: _____ Z (shift + Z or capital Z)

* Zoom out: _____ z (lower case z)

* Zoom to fit: _____ F

* Previous view: _____ Ctrl+Shift+Z

View Orientation

* View Orientation Menu: _____ Space bar

* Front: _____ Ctrl+1

* Back: _____ Ctrl+2

* Left: _____ Ctrl+3

* Right: _____ Ctrl+4

* Top: _____ Ctrl+5

* Bottom: _____ Ctrl+6

* Isometric: _____ Ctrl+7

Selection Filter & Misc.

* Filter Edges: _____ e

* Filter Vertices: _____ v

* Filter Faces: _____ x

* Toggle Selection filter toolbar: _____ F5

* Toggle Selection Filter toolbar (on/off): _____ F6

* New SOLIDWORKS document: _____ F1

* Open Document: _____ Ctrl+O

* Open from Web folder: _____ Ctrl+W

* Save: _____ Ctrl+S

* Print: _____ Ctrl+P

* Magnifying Glass Zoom _____ g

* Switch between the SOLIDWORKS documents _____ Ctrl + Tab

SOLIDWORKS Sample Customized Hot Keys

Function Keys

F1	SW-Help
F2	2D Sketch
F3	3D Sketch
F4	Modify
F5	Selection Filters
F6	Move (2D Sketch)
F7	Rotate (2D Sketch)
F8	Measure
F9	Extrude
F10	Revolve
F11	Sweep
F12	Loft

Sketch

C	Circle
P	Polygon
E	Ellipse
O	Offset Entities
Alt + C	Convert Entities
M	Mirror
Alt + M	Dynamic Mirror
Alt + F	Sketch Fillet
T	Trim
Alt + X	Extend
D	Smart Dimension
Alt + R	Add Relation
Alt + P	Plane
Control + F	Fully Define Sketch
Control + Q	Exit Sketch